Advances in Mechanics and Mathematics

Volume 38

More information about this series at http://www.springer.com/series/5613

Brian Straughan

Mathematical Aspects of Multi–Porosity Continua

Springer

Brian Straughan
Department of Mathematical Sciences
University of Durham
Durham, UK

ISSN 1571-8689 ISSN 1876-9896 (electronic)
Advances in Mechanics and Mathematics
ISBN 978-3-319-88895-8 ISBN 978-3-319-70172-1 (eBook)
https://doi.org/10.1007/978-3-319-70172-1

Mathematics Subject Classification (2010): 74B05, 74B20, 74H25, 74H35, 74L10, 74E99, 35A21, 35B30

Printed on acid-free paper

This Springer imprint is published by Springer Nature
The registered company is Springer International Publishing AG
The registered company address is: Gewerbestrasse 11, 6330 Cham, Switzerland

Preface

This book is devoted to describing theories for porous media where such pores have an inbuilt macro structure and a micro structure. For example, a double porosity material has pores on a macro scale, but additionally there are cracks or fissures in the solid skeleton. The actual body is allowed to deform and thus the underlying theory is one of elasticity. Various different descriptions are reviewed.

Mathematical analyses of double and triple porosity materials are included concentrating on uniqueness and stability studies in chapters 5 to 7. In chapters 8 and 9 the emphasis is on wave motion in double porosity materials with special attention paid to nonlinear waves.

The final chapter embraces a novel area where an elastic body with a double porosity structure is analysed, but the thermodynamics allows for heat to travel as a wave rather than simply by diffusion.

I would like to thank Professor Merab Svanadze for many helpful discussions on multi - porosity elastic materials. I am also indebted to four anonymous referees whose comments and suggestions have led to substantial improvements in the book. In addition, I should like to thank Donna Chernyk for her help with editorial matters.

Durham *Brian Straughan*

Contents

Chapter 1
Introduction

1.1 Multiple Porosity and Applications

This book is dedicated to describing theories and analysis for the class of elastic materials which display a multiple porosity structure. If one has an elastic body which has pores (holes) in it then one begins with the macro porosity to which there is an associated pressure in the pores. This gives rise to the traditional concept of porosity in a porous material, which is that the porosity of a sample is the volume of fluid in the sample divided by the total volume of the sample, namely that comprised of the solid skeleton and the fluid content. In a multiple porosity elastic body as we are focussing on in this book, we allow additional porosities which are due to the microstructure of the skeleton. We begin with a double porosity elastic material which has macro pores in the body but in addition there is a micro porosity which arises because of fissures or cracks in the solid skeleton and we here assume that associated to this micro structure there is an independent pressure field in the fluid saturating the micro pores. An excellent example of a micro porosity - macro porosity situation may be seen in the photographs in Masin *et al.* [158]. These photographs display a pile of soil comprised of large pieces of clay which have been dug from open-cast mines and they are now in landfills. The larger pieces of clay form a macro porosity body but they additionally contain many fissures (or cracks), and the pictures show the degradation of the macro porosity over a period of ten years which eventually leaves a pile of finer material which contains a porosity due to the micro porous structure. We point out below that double porosity materials occur frequently in real life and we give specific applications of such.

Figure 1.1 shows a double porosity material which could be constructed, for example, by using small spherical glass beads. The next figure 1.2 shows a similar double porosity structure with a less regular geometry. In figure 1.3 we show a double porosity structure in a wall. The large stones lead to a macro porosity but each stone itself has many pores which give the micro porosity. Figure 1.4 is a different view of the same wall.

B. Straughan, *Mathematical Aspects of Multi–Porosity Continua*, Advances in Mechanics and Mathematics 38, https://doi.org/10.1007/978-3-319-70172-1_1

The double porosity medium in figure 1.5 is created by putting together many pieces of lava from Mount Etna. The saturating fluid is water. Figure 1.6 is another view of the water saturated lava double porosity medium. The two figures in figure 1.7 display a possible man made double porosity elastic medium. The spheres could be made of glass, rubber, or many things. We could have many such spheres and the macro porosity is given by the pores between the spheres themselves while the micro porosity arises because of the cylindrical holes in the spheres. One could create many similar double porosity media by varying the way the holes are drilled and by changing the saturating fluid or gas.

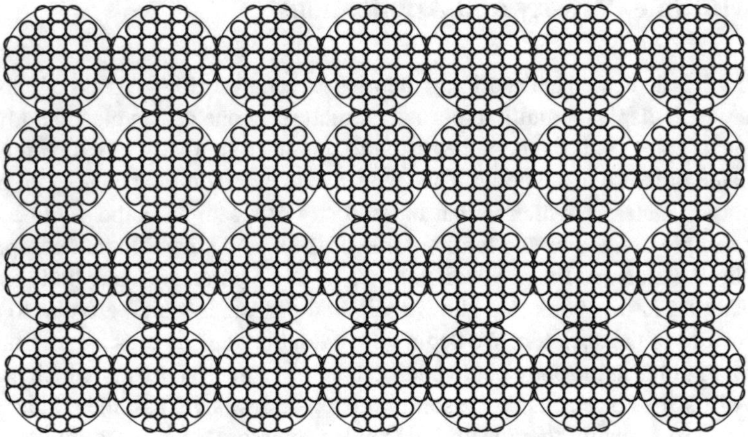

Fig. 1.1 An example of a double porosity medium. The micro porosity is defined by the smaller spheres whereas the macro porosity defined by the larger "fictitious" spheres.

Double porosity elastic materials are very important but we should point out that recent research literature is also concentrating on triple porosity elastic materials, or even bodies which contain a quadruple porosity structure, or perhaps even a quintuple porosity structure, cf. Aguilera & Aguilera [1], Bai *et al.* [9], Bai & Roegiers [10], Debois *et al.* [65], Deng *et al.* [69], He *et al.* [103], Kuznetsov & Nield [148], Olusola *et al.* [169], Solano *et al.* [198], Straughan [208, 209], Svanadze [215]. In the case of an elastic material with triple porosity the body will possess three levels of pore structures. The first is the largest visible porosity known as macro porosity, the second represents an intermediate case which is known as meso porosity, and the final scenario is referred to as a micro porosity. In the case of quadruple and quintuple porosity there is a hierarchical structure of porosity, each one being on a successively smaller scale than the former. In the quadruple porosity case one has a

Fig. 1.2 This is a schematic picture of a double porosity or bidispersive material. Note the large pore structure in between the darker elements which are themselves composed of small spherical bead like bodies giving a dual porosity raspberry-like structure.

macro porosity, then a meso porosity, then a micro porosity, and finally a sub-micro porosity. For the quintuple case the hiearchy goes in the order macro pores, meso pores, micro pores, sub-micro pores, and then nano pores. The article by Solano *et al.* [198] observes that the porosity scales go from Km, m, mm, μm to nm. In the chemical engineering industry the nano pore effects are likely to play a very important role, see e.g. Enterria *et al.* [73], Huang *et al.* [107], Ly *et al.* [156] and Said *et al.* [184].

The need for theories for multiple porosity elasticity and the associated mathematical, physical and numerical analysis which accompanies such theories is undoubtedly driven by the myriad of applications which exist and which are coming to light continuously. We do not attempt a thorough review of all the application areas but we believe it is worth specifically drawing attention to some.

The first application area is in mathematical biology and the associated field of health. In this category we mention the use of biomaterials in clinical situations, Dejaco *et al.* [67], the design of membrane - based bioartifical livers, Dufresne *et al.* [70], interstitial fluid flow in bones together with the sector of bone replacement and recovery technology, see e.g. Svanadze & Scalia [216, 217], Sakamoto & Matsumoto [185], Zhou *et al.* [232], and tissue engineering strategy for bone defects, Dejaco *et al.* [66]. Replacement of damaged long bones in human beings is a major

Fig. 1.3 This is a picture of a double porosity material. The stones themselves have many small pores in them. Picture taken near the Necropoli di Pantalica, Sicily.

problem for a surgeon since the porosity of the bone can vary from 14 % in the outer layer bone to 52 % in the inner layer. Indeed to adequately model a long bone one may require a multi - porosity theory which is applicable to a graded porosity material, see, e.g. Zhou *et al.* [232].

Another very important area of application for multiple porosity elasticity is in geophysics. For example a careful description of landslides may well require employment of double porosity theory, see e.g. Montrasio *et al.* [162], Borja *et al.* [24], Borja & White [25], Pooley [174]. Much data has been collected and analysed regarding landslides in the double porosity system in the Vesuvian and Phlegrean deposits in the neighbourhood of the city of Naples. The porosity is greater than seventy per cent for Vesuvian deposits and lower for Phlegrean ones, see Scotto di

Fig. 1.4 This is a picture of a double porosity material. The stones themselves have many small pores in them. Picture taken near the Necropoli di Pantalica, Sicily.

Santolo & Evangelista [187]. A study of carbon sequestration in fissured aquifers likewise may require porosity at a multiple scale, see Carneiro [30].

An application area which is frequently in the news is that concerned with energy production. Into this category we mention nuclear waste treatment, see, Said *et al.* [184], nuclear material behaviour, see Spicer *et al.* [200], fuel cell technology, see e.g. Yuan & Sundén [229], recovery of methane from a coal bed, see e.g. Wei & Zhang [226], Zou *et al.* [233], oil reservoir recovery, see e.g. Aguilera & Aguilera [1], Bai *et al.* [9], Bai & Roegiers [10], Deng *et al.* [69], Olusola *et al.* [169], and Zhao *et al.* [231]. Finally in this category we list modelling gas production via the controversial technique of hydraulic fracturing (which is frequently referred to as "fracking"), cf. Sarma & Aziz [188] and Kim & Moridis [136].

Fig. 1.5 This is a picture of a double porosity material. The large stones are lava from Mount Etna. The saturating fluid is water.

The theory of sound waves in porous bodies is a controversial area and there are many empirical models for finding wavespeeds and other effects such as attenuation, together with several rigorous articles on acoustics. However, acoustics, and especially sound propagation in porous media, is a highly active area, see e.g. Capelli *et al.* [27], Christov [46], Christov *et al.* [47], Christov *et al.* [49], Dazel *et al.* [62], Duwairi [71], Fellah *et al.* [81], Gorgas *et al.* [95], Jiang *et al.* [118], Jordan [122, 123, 124, 125], Jordan & Fulford [126], Jordan *et al.* [121], Kaltenbacher [132], Molinari *et al.* [161], Rossmanith & Puri [181, 180], Sgard *et al.* [192], Venegas & Umnova [223], Winkler & Murphy [227]. The theory of poroacoustics in an elastic body with a multi-porous structure is one which is undoubtedly rich in future applications.

Fig. 1.6 This is a picture of a double porosity material. The large stones are lava from Mount Etna. The saturating fluid is water.

Finally we mention the application area of other consumables such as provision of clean drinking water. This may be from a carbonate aquifer, see e.g. Zuber & Motyka [234], Ghasemizadeh *et al.* [91]. Providing uncontaminated drinking water is vital to the human race and contaminants may even be carcinogenic and double porosity effects on contamination in aquifers are being analysed, see e.g. Fitzpatrick [83]. It is of interest to observe that capacitive deionization is an electrochemical water treatment process which may be a viable alternative to treat water in an energy saving manner and this process may employ dual porosity electrodes, see Gabitto & Tsouris [87].

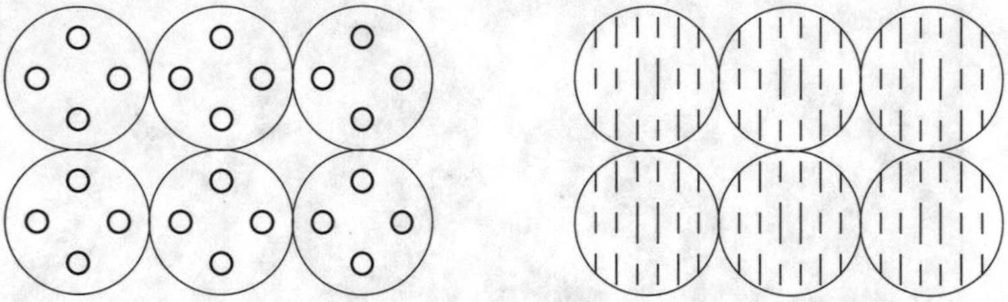

Fig. 1.7 An example of a man made double porosity medium consisting of twelve spheres. The pattern can be continued by adding more spheres appropriately. The micro porosity is defined by the cylindrical holes in the spheres whereas the macro porosity is defined by the gaps between the larger spheres. The spheres to the right represent a view from the side of the material whereas the spheres to the left represent a view from the top.

1.2 Notation

Standard indicial notation is used throughout this book together with the Einstein summation convention for repeated indices. Standard vector or tensor notation is also employed where appropriate. For example, for a function u or a vector function u_i we write

$$u_x \equiv \frac{\partial u}{\partial x} \equiv u_{,x} \qquad u_{i,t} \equiv \frac{\partial u_i}{\partial t} \qquad u_{i,i} \equiv \frac{\partial u_i}{\partial x_i} \equiv \sum_{i=1}^{3} \frac{\partial u_i}{\partial x_i}$$

$$u_{j,ij} \equiv \frac{\partial^2 u_j}{\partial x_i \partial x_j} \equiv \sum_{j=1}^{3} \frac{\partial^2 u_j}{\partial x_i \partial x_j}, \qquad i = 1,2 \text{ or } 3,$$

$$\Delta u \equiv u_{,ii} \equiv \frac{\partial^2 u}{\partial x_i \partial x_i} \equiv \sum_{j=1}^{3} \frac{\partial^2 u}{\partial x_i \partial x_i},$$

$$u_j u_{i,j} \equiv u_j \frac{\partial u_i}{\partial x_j} \equiv \sum_{j=1}^{3} u_j \frac{\partial u_i}{\partial x_j}, \qquad i = 1,2 \text{ or } 3,$$

$$(a_{ijkh} u_{k,h})_{,j} = \frac{\partial}{\partial x_j}\left(a_{ijkh} \frac{\partial u_k}{\partial x_h}\right) \equiv \sum_{j,k,h=1}^{3} \frac{\partial}{\partial x_j}\left(a_{ijkh} \frac{\partial u_k}{\partial x_h}\right), \qquad i = 1,2 \text{ or } 3,$$

where Δ denotes the Laplace operator, namely in three dimensions

$$\Delta u = \frac{\partial^2 u}{\partial x_1^2} + \frac{\partial^2 u}{\partial x_2^2} + \frac{\partial^2 u}{\partial x_3^2} = \frac{\partial^2 u}{\partial x^2} + \frac{\partial^2 u}{\partial y^2} + \frac{\partial^2 u}{\partial z^2}.$$

In the case where a repeated index sums over a range different from 1 to 3 this will be pointed out in the text. Note that in bold face notation

$$u_{i,i} \equiv \operatorname{div} \mathbf{u}, \qquad \Delta u_i \equiv \Delta \mathbf{u} \qquad \text{and} \qquad u_j u_{i,j} \equiv (\mathbf{u} \cdot \nabla)\mathbf{u}.$$

As indicated above, a subscript t denotes partial differentiation with respect to time. When a superposed dot is used it also means the partial derivative with respect to time, e.g.

$$\dot{u}_i \equiv \frac{\partial u_i}{\partial t},$$

or

$$\dot{p} = \frac{\partial p}{\partial t},$$

where $u_i(\mathbf{x},t)$ will typically denote the elastic displacement field and $p(\mathbf{x},t)$ is another field such as pressure.

To further clarify the indicial notation used, we note for example, the product of two second order tensors A_{ij} and B_{ij}, say $\mathbf{C} = \mathbf{AB}$, may be written

$$\begin{aligned} C_{ij} &= A_{ik} B_{kj} \\ &\equiv \sum_{k-1}^{3} A_{ik} B_{kj}. \end{aligned}$$

Further, if $\mathbf{x}$ and $\mathbf{b}$ are vectors then $\mathbf{Ax} = \mathbf{b}$ may be written as

$$\begin{aligned} b_i &= A_{ik} x_k \\ &\equiv \sum_{k=1}^{3} A_{ik} x_k. \end{aligned}$$

We now derive some useful expressions. Recall the definitions of the Kronecker delta, δ_{ij}, and the alternating tensor, ε_{ijk}. Namely,

$$\delta_{ij} = \begin{cases} 1, & \text{if } i = j; \\ 0, & \text{if } i \neq j, \end{cases}$$

i.e. $\delta_{11} = \delta_{22} = \delta_{33} = 1$, the rest are zero. Also,

$$\begin{aligned} \varepsilon_{123} = \varepsilon_{231} = \varepsilon_{312} = +1, \\ \varepsilon_{213} = \varepsilon_{321} = \varepsilon_{132} = -1, \end{aligned} \tag{1.1}$$

the rest are zero.

In terms of the Kronecker delta and the alternating tensor,

$$\varepsilon_{ijk}\varepsilon_{irs} = \delta_{jr}\delta_{ks} - \delta_{js}\delta_{kr}$$

and then for a vector field $\mathbf{A}$,

$$(\text{curl}\,\mathbf{A})_i = \varepsilon_{ijk}A_{k,j}$$

and

$$\begin{aligned}(\text{curl}\,\text{curl}\,\mathbf{A})_i =& \varepsilon_{ijk}\varepsilon_{krs}A_{s,rj}\\ =& \varepsilon_{kij}\varepsilon_{krs}A_{s,rj}\\ =& (\delta_{ir}\delta_{js} - \delta_{is}\delta_{jr})A_{s,rj}\\ =& A_{j,ij} - \Delta A_i\,. \end{aligned} \tag{1.2}$$

Thus, if v_i is a divergence free vector field, $v_{i,i} = 0$, and then

$$(\text{curl}\,\text{curl}\,\mathbf{v})_i = -\Delta v_i\,,$$

where $\Delta = \partial^2/\partial x^2 + \partial^2/\partial y^2 + \partial^2/\partial z^2$ is the Laplace operator seen earlier.

The vector product (or cross product) of two vectors $\mathbf{A}$ and $\mathbf{B}$ may be written in component form in indicial notation as

$$(\mathbf{A}\times\mathbf{B})_i = \varepsilon_{ijk}A_jB_k\,.$$

We sometimes have recourse to deal with diagonal matrices and so we introduce the notation

$$\mathbf{A} = \text{diag}\,(a_1, a_2, a_3) \tag{1.3}$$

to mean $\mathbf{A}$ is the diagonal matrix

$$\mathbf{A} = \begin{pmatrix} a_1 & 0 & 0\\ 0 & a_2 & 0\\ 0 & 0 & a_3 \end{pmatrix}.$$

The letter Ω will denote a fixed, bounded region of 3-space with boundary, Γ, sufficiently smooth to allow applications of the divergence theorem.

The symbols $\|\cdot\|$ and $(\cdot,\cdot)$ will denote, respectively, the L^2 norm on Ω, and the inner product on $L^2(\Omega)$,

$$\int_\Omega f^2 dx = \|f\|^2 \quad \text{and} \quad (f,g) = \int_\Omega fg\,dx.$$

We additionally denote integration of a function over Ω by $< \cdot >$, e.g.

$$\int_\Omega f\,dx = < f >\,.$$

We sometimes have recourse to use the norm on $L^p(\Omega)$, $1 < p < \infty$, and then we write

$$\|f\|_p = \left(\int_\Omega |f|^p dx \right)^{1/p} .$$

For a second order tensor α_{ij} we say that α_{ij} is symmetric if $\alpha_{ij} = \alpha_{ji}$, and it is positive-definite if

$$\alpha_{ij}\xi_i\xi_j \geq \alpha_0\xi_i\xi_i, \qquad \forall\, \xi_i,$$

for some constant $\alpha_0 > 0$. Further α_{ij} is *non-negative* if

$$\alpha_{ij}\xi_i\xi_j \geq 0, \qquad \forall\, \xi_i.$$

For a fourth order tensor a_{ijkh} we say a_{ijkh} is *positive-definite* if

$$a_{ijkh}\xi_{ij}\xi_{kh} \geq a_0\xi_{ij}\xi_{ij}, \qquad \forall\, \xi_{ij},$$

for some $a_0 > 0$. We refer to a_{ijkh} as being non-negative if

$$a_{ijkh}\xi_{ij}\xi_{kh} \geq 0, \qquad \forall\, \xi_{ij}.$$

We frequently deal with the expression

$$(a_{ijkh}u_{k,h})_{,j} = \frac{\partial}{\partial x_j}\left(a_{ijkh}\frac{\partial u_k}{\partial x_h} \right), \tag{1.4}$$

where, in general, $a_{ijkh} = a_{ijkh}(\mathbf{x})$, and $i = 1,2,3$ in three-dimensions, but sometimes $i = 1,2$ if we restrict attention to the two-dimensional case. Note that (1.4) employs the Einstein summation convention so it is

$$(a_{ijkh}u_{k,h})_{,j} = \sum_{h=1}^{3}\sum_{k=1}^{3}\sum_{j=1}^{3} \frac{\partial}{\partial x_j}\left(a_{ijkh}\frac{\partial u_k}{\partial x_h} \right),$$

where $i = 1,2$ or 3, in three-dimensions.

To be specific about what (1.4) involves using the summation convention we write out terms explicitly in the two-dimensional case, for a_{ijkh} constants. Suppose also a_{ijkh} satisfy the symmetries

$$a_{ijkh} = a_{khij} = a_{jikh}. \tag{1.5}$$

Since in two-dimensions i, j, k and h each take values 1 and 2 we find there are 16 individual coefficients for a_{ijkh}. However, the symmetries (1.5) reduce this number to 6. These may be seen to be, where we introduce $A, \dots, F$ as indicated

$$\begin{aligned} &a_{1111} = A, \qquad a_{2222} = B, \qquad a_{1112} = a_{1211} = a_{2111} = a_{1121} = C,\\ &a_{1122} = a_{2211} = D, \qquad a_{1221} = a_{1212} = a_{2121} = a_{2112} = E,\\ &a_{1222} = a_{2221} = a_{2212} = a_{2122} = F. \end{aligned}$$

With a_{ijkh} constants, expression (1.4) may be written as, for $i = 1,2$,

$$a_{1jkh}u_{k,hj} = Au_{xx} + 2Cu_{xy} + Eu_{yy} + Cv_{xx} + (D+E)v_{xy} + Fv_{yy}\,,$$
$$a_{2jkh}u_{k,hj} = Bv_{yy} + 2Fv_{xy} + Ev_{xx} + Cu_{xx} + (D+E)u_{xy} + Fu_{yy}\,,$$

where we have written $\mathbf{u} = (u,v) = (u_1,u_2)$ and $u_{xx} = u_{,xx} = \partial^2 u/\partial x^2$, etc.

Throughout the book we make frequent use of inequalities. In particular, we often use the Cauchy-Schwarz inequality for two functions f and g, i.e.

$$\int_\Omega fg\,dx \le \left(\int_\Omega f^2 dx\right)^{1/2} \left(\int_\Omega g^2 dx\right)^{1/2}, \tag{1.6}$$

or what is the same in L^2 norm and inner product notation,

$$(f,g) \le \|f\|\,\|g\|. \tag{1.7}$$

The arithmetic-geometric mean inequality (with a constant weight $\alpha > 0$) is, for $a,b \in \mathbb{R}$,

$$ab \le \frac{1}{2\alpha}a^2 + \frac{\alpha}{2}b^2, \tag{1.8}$$

and this is easily seen to hold since

$$\left(\frac{a}{\sqrt{\alpha}} - \sqrt{\alpha}b\right)^2 \ge 0.$$

Another inequality we frequently have recourse to is Young's inequality, which for $a,b \in \mathbb{R}$ we may write as

$$ab \le \frac{|a|^p}{p} + \frac{|b|^q}{q}, \qquad \frac{1}{p} + \frac{1}{q} = 1, \quad p,q \ge 1. \tag{1.9}$$

For Ω a bounded domain in $\mathbb{R}^3$ and a function $u = 0$ on Γ (the boundary of Ω) we shall employ the Poincaré inequality, namely

$$\lambda_1 \|u\|^2 \le \|\nabla u\|^2 \tag{1.10}$$

where $\lambda_1 > 0$ is a constant which depends on Ω. Inequality (1.10) arises from solutions to the membrane problem,

$$\begin{aligned} \Delta u + \lambda u &= 0, \qquad \text{in } \Omega, \\ u &= 0, \qquad \text{on } \Gamma, \end{aligned} \tag{1.11}$$

cf. for example, Payne [170], Payne & Weinberger [173], Payne *et al.* [171], Gilbarg & Trudinger [92], sections 7.7, 7.8.

For Ω sufficiently regular it can be shown that there are a countably infinite number of eigenvalues which may be ordered so that

$$0 < \lambda_1 \le \lambda_2 \le \ldots$$

The first eigenvalue is defined by the characterisation

$$\lambda_1 = \inf_{u \in H_0^1(\Omega)} \frac{\int_\Omega |\nabla u|^2 dx}{\int_\Omega u^2 dx} = \inf_{u \in H_0^1(\Omega)} \frac{\|\nabla u\|^2}{\|u\|^2} .$$

In this definition $H_0^1(\Omega)$ is the Sobolev space of $L^2(\Omega)$ functions whose gradient is in $L^2(\Omega)$ and which have compact support in Ω.

1.3 Classical Elastodynamics

The equations of classical linear elastodynamics are described in detail in chapter 2 of Knops & Payne [143] and in pages 723 – 727 of Truesdell & Toupin [221]. A derivation of these equations follows as in section 3.2.2 by assuming the temperature θ to be constant. The relevant equations are then (3.34) with $\theta \equiv$constant. We simply present these equations at this stage. If $u_i(\mathbf{x},t)$ is the displacement of a point in a three-dimensional elastic body then the general form for the equations of motion is

$$\rho \ddot{u}_i = \frac{\partial t_{ji}}{\partial x_j} + \rho f_i \tag{1.12}$$

where $\rho(\mathbf{x},t)$ is the density of the body, f_i is a prescribed body force, $\mathbf{x}$ is a point in the body, and t_{ji} is the stress tensor.

The general form for the stress tensor for a linearly anisotropic elastic solid is

$$t_{ij} = a_{ijkh} \frac{\partial u_k}{\partial x_h} . \tag{1.13}$$

Since i, j, k and h each take values 1,2 and 3 it follows that in the general case the elastic coefficients a_{ijkh} amount to 81 in number.

Symmetry conditions usually dictate that the stress tensor t_{ij} is symmetric and then we denote it by σ_{ij}, so that

$$\sigma_{ij} = \sigma_{ji} .$$

Let $e_{ij} = (u_{i,j} + u_{j,i})/2$ be the strain tensor and then for *Cauchy elasticity*, see Truesdell & Toupin [221], section 301, equation (1.13) may be replaced by

$$\sigma_{ij} = a_{ijkh} e_{kh} . \tag{1.14}$$

In this case a_{ijkh} satisfies the symmetry relations

$$a_{ijkh} = a_{jikh} = a_{ijhk} . \tag{1.15}$$

There are no longer 81 independent elastic coefficients a_{ijkh}, since this number reduces to 36, see exercise 1.3.

A more restricted class of material is known as *Green elasticity*, see Truesdell & Toupin [221], section 301, Knops & Payne [143], p. 12, and for this one replaces conditions (1.15) by

$$a_{ijkh} = a_{khij} = a_{jikh}\,. \tag{1.16}$$

With the symmetries of (1.16) the number of independent elastic coefficients is reduced to 21, see exercise 1.3. In this book we shall almost exclusively restrict attention to employing equations (1.16).

For Green elasticity the evolutionary equations (1.12) reduce to

$$\rho\frac{\partial^2 u_i}{\partial t^2} = \frac{\partial}{\partial x_j}\Big(a_{ijkh}\frac{\partial u_k}{\partial x_h}\Big) + \rho f_i\,, \tag{1.17}$$

where the elastic coefficients a_{ijkh} satisfy the symmetry conditions (1.16).

In this book we shall often require only conditions (1.16). However, in addition it is sometimes necessary to be more restrictive and impose further constraints on the elastic coefficients. We shall say a_{ijkh} are *positive - definite* if

$$a_{ijkh}\xi_{ij}\xi_{kh} \geq a_0\xi_{ij}\xi_{ij}\,, \tag{1.18}$$

for a_0 a positive constant, for all tensors ξ_{ij}, where it is understood that (1.18) holds for all points in the elastic body under consideration.

The elastic coefficients are *positive semi - definite* (sometimes referred to as simply *positive* or *non-negative*) if

$$a_{ijkh}\xi_{ij}\xi_{kh} \geq 0\,, \tag{1.19}$$

$\forall\,\xi_{ij}$, $\mathbf{x}\in\mathscr{B}$, where $\mathscr{B}$ is the domain of the elastic body.

For a connected body $\mathscr{B}$ we say the elastic coefficients satisfy the *strong ellipticity* condition when

$$a_{ijkh}\eta_i\eta_k\xi_j\xi_h > 0, \qquad \forall\,\xi_i,\eta_i \neq 0. \tag{1.20}$$

1.3.1 Isotropic Linear Elasticity

In this case the material response in the body is the same in all directions and this condition is met by requiring that the elastic coefficients are given by

$$a_{ijkh} = \lambda\delta_{ij}\delta_{kh} + \mu(\delta_{ik}\delta_{jh} + \delta_{ih}\delta_{jk}).$$

The coefficients λ and μ are the Lamé and shear moduli, respectively, and they are related to Poisson's ratio σ, and to Young's modulus E, by

$$\lambda = \frac{2\mu\sigma}{1-2\sigma}\,, \qquad \mu = \frac{E}{2(1+\sigma)}\,, \tag{1.21}$$

see e.g. Knops & Payne [143], p. 10.

The stress relation (1.14) reduces in the isotropic case to

$$\sigma_{ij} = \lambda e_{rr}\delta_{ij} + 2\mu e_{ij}.$$

This then gives rise to the dynamical equations for an isotropic linear elastic body as

$$\rho \ddot{u}_i = \mu \Delta u_i + (\lambda + \mu) u_{j,ij} + \rho f_i. \tag{1.22}$$

If one requires positive-definiteness of the elastic coefficients then λ and μ are required to satisfy, see Knops & Payne [143], p. 19,

$$3\lambda + 2\mu > 0 \qquad \text{and} \qquad \mu > 0. \tag{1.23}$$

This is equivalent to the conditions

$$-1 < \sigma < \frac{1}{2}, \qquad \mu > 0.$$

When the strong ellipticity condition is required the λ and μ satisfy

$$\mu(\lambda + 2\mu) > 0. \tag{1.24}$$

see e.g. Knops & Payne [143], p. 21, Chirita & Ghiba [44]. This condition is equivalent to requiring

$$-1 < \sigma < \frac{1}{2}, \qquad 1 < \sigma < \infty, \qquad \mu \neq 0.$$

There are many crystal classes which reduce the number of independent elastic coefficients from 21 in number, in addition to the isotropic case, cf. Chirita *et al.* [43]. We here consider two such classes, namely, those of cubic symmetry and for transverse isotropy.

1.3.2 Cubic Crystal Class

We begin with the cubic crystal class for which there are three independent elastic coefficients. It is convenient to introduce the standard notation

$$C_{ij} = a_{iijj}, \qquad i = 1, 2, 3,$$

although one also defines all components of the 6×6 matrix (C_{ij}). In terms of the coefficients a_{ijkh} the components of C_{ij} are given in full in equations (1.3) of Chirita & Danescu [42]. Further information is contained in the article of Mouhat & Coudert [163].

For a cubic material the three independent elastic coefficients are given by

$$
\begin{aligned}
C_{11} &= C_{22} = C_{33}\,,\\
C_{12} &= C_{23} = C_{31}\,,\\
C_{44} &= C_{55} = C_{66}\,,
\end{aligned}
\tag{1.25}
$$

where

$$
\begin{aligned}
C_{11} &= a_{1111}\,, & C_{22} &= a_{2222}\,, & C_{33} &= a_{3333}\,,\\
C_{12} &= a_{1122}\,, & C_{23} &= a_{2233}\,, & C_{31} &= a_{3311}\,,\\
C_{44} &= a_{2323}\,, & C_{55} &= a_{1313}\,, & C_{66} &= a_{1212}\,,
\end{aligned}
\tag{1.26}
$$

with

$$
\begin{aligned}
a_{1123} = a_{1131} = a_{1112} = a_{2223} = a_{2231} = a_{2212} = 0,\\
a_{3323} = a_{3331} = a_{3312} = a_{2331} = a_{2312} = a_{3112} = 0,
\end{aligned}
$$

where the Green symmetries (1.16) still hold. The matrix **C** may be written as, see e.g. Mouhat & Coudert [163],

$$
\mathbf{C} = \begin{pmatrix}
C_{11} & C_{12} & C_{12} & 0 & 0 & 0\\
C_{12} & C_{11} & C_{12} & 0 & 0 & 0\\
C_{12} & C_{12} & C_{11} & 0 & 0 & 0\\
0 & 0 & 0 & C_{44} & 0 & 0\\
0 & 0 & 0 & 0 & C_{44} & 0\\
0 & 0 & 0 & 0 & 0 & C_{44}
\end{pmatrix}
$$

The governing equations of elastodynamics are still (1.17).

1.3.3 Transversely Isotropic Materials

A transversely isotropic material is one where there is a preferred direction and the material is isotropic in all directions orthogonal to the preferred one. In the case of a transversely isotropic material there are five independent elastic coefficients, cf. Chirita [37]. When the direction of the transverse isotropy is the x_3−axis the non-zero elastic coefficients are given by

$$
\begin{aligned}
&C_{11} = C_{22}\,, \quad C_{23} = C_{13}\,, \qquad C_{33}\,,\\
&C_{44} = C_{55} = a_{2323} = a_{3131}\,, \qquad C_{66} = a_{1212} = \frac{1}{2}(C_{11} - C_{12})\,.
\end{aligned}
\tag{1.27}
$$

Again, equations (1.16) and (1.17) still hold. The matrix **C** is in this case, cf. Mouhat & Coudert [163],

$$\mathbf{C} = \begin{pmatrix} C_{11} & C_{12} & C_{13} & 0 & 0 & 0 \\ C_{12} & C_{11} & C_{13} & 0 & 0 & 0 \\ C_{13} & C_{13} & C_{33} & 0 & 0 & 0 \\ 0 & 0 & 0 & C_{44} & 0 & 0 \\ 0 & 0 & 0 & 0 & C_{44} & 0 \\ 0 & 0 & 0 & 0 & 0 & C_{66} \end{pmatrix}$$

1.3.4 Uniqueness by the Energy Method

In this section we consider the boundary-initial value problem $\mathscr{P}$ given by equations (1.17) in $\Omega \times (0,T)$ for some $T > 0$, together with the boundary conditions

$$u_i(\mathbf{x},t) = u_i^B(\mathbf{x}), \qquad \text{on } \Gamma \times (0,T], \tag{1.28}$$

and the initial conditions

$$u_i(\mathbf{x},0) = v_i(\mathbf{x}), \quad \dot{u}_i(\mathbf{x},0) = w_i(\mathbf{x}), \qquad \mathbf{x} \in \Omega, \tag{1.29}$$

where u_I^B, v_i and w_i are prescribed functions. We suppose the elastic coefficients satisfy the symmetries (1.16) and are also positive semi-definite in the sense of (1.19).

To establish uniqueness of a solution to $\mathscr{P}$ we suppose there are two solutions u_i^1 and u_i^2 which satisfy $\mathscr{P}$ for the same boundary data function u_i^B, for the same body force f_i, and for the same initial data functions v_i and w_i. Then introduce the difference solution u_i by

$$u_i = u_i^1 - u_i^2 .$$

By inspection one sees that $u_i(\mathbf{x},t)$ satsifies the boundary-initial value problem given by

$$\begin{aligned} &\rho \ddot{u}_i = (a_{ijkh} u_{k,h})_{,j}, && \text{in } \Omega \times (0,T), \\ &u_i(\mathbf{x},t) = 0, && \text{on } \Gamma \times (0,T], \\ &u_i(\mathbf{x},0) = 0, && \dot{u}(\mathbf{x},0) = 0, \qquad \mathbf{x} \in \Omega. \end{aligned} \tag{1.30}$$

Next, mulitply $(1.30)_1$ by $\dot{u}_i$ and integrate over Ω to find after some integration by parts and use of $(1.30)_2$,

$$\frac{dE}{dt} = 0, \tag{1.31}$$

where

$$E(t) = \frac{1}{2} < \rho \dot{u}_i \dot{u}_i > + \frac{1}{2} < a_{ijkh} u_{i,j} u_{k,h} > . \tag{1.32}$$

Then, integrating (1.31) we obtain

$$E(t) = E(0) = 0, \tag{1.33}$$

where $E(0) = 0$ follows from $(1.30)_3$ and the definition of E. Now employ inequality (1.19) to see that

$$0 \leq < \rho \dot{u}_i \dot{u}_i > \leq 0, \qquad t \in (0,T),$$

and so $\dot{u}_i \equiv 0$ on $(0,T)$. Since $u_i(\mathbf{x},0) = 0$ it therefore follows that $u_i \equiv 0$ for $\mathbf{x} \in \Omega$, $t \in [0,T)$ and uniqueness of a solution to $\mathscr{P}$ is established.

1.3.5 Uniqueness for an Unbounded Domain

We now wish to consider the analogous uniqueness question to that of section 1.3.4 but now where Ω is the region exterior to a bounded region $\Omega_0 \subset \mathbb{R}^3$, and Γ is the boundary of Ω_0, i.e. Γ is the interior boundary of Ω. We shall assume that the origin lies inside Ω_0.

We wish to allow the solution to grow as $|\mathbf{x}| \to \infty$ and thus one cannot simply employ a straightforward energy argument as in section 1.3.4 since this would require strong decay of the solution as $|\mathbf{x}| \to \infty$. Thus, we employ a weighted energy method due to Rionero & Galdi [178]. Hence, we introduce the weight function $g(r)$, $r = \sqrt{x_i x_i}$, by the equation

$$g(r) = e^{-\delta r}, \tag{1.34}$$

for δ a constant we select later. Better uniqueness results than those established here may be achieved with a more judicious (and more complicated) choice of weight as shown in Galdi & Rionero [88].

We now proceed as in section 1.3.4 and introduce the difference solution $u_i = u_i^1 - u_i^2$ where u_i again satisfies (1.30) but now Ω is an exterior domain. We suppose the elastic coefficients satisfy the positive - definiteness condition (1.18) for a constant $a_0 > 0$ and the symmetry conditions (1.16). In addition we suppose that

$$|a_{ijkh}| \leq M \qquad \text{and} \qquad |\rho(\mathbf{x})| \geq \rho_0 > 0, \tag{1.35}$$

for all $\mathbf{x} \in \Omega$ and where M, ρ_0 are positive constants. The class of solutions we allow is those where

$$|\dot{u}_i|, |u_{k,h}| \leq k e^{\lambda r}, \tag{1.36}$$

for some positive constants k, λ.

We now multiply equation $(1.30)_1$ by $g\dot{u}_i$ and integrate over Ω. We suppose now that δ is chosen so that

$$\delta > 2\lambda\,. \tag{1.37}$$

We have to integrate by parts in the resulting term $\int_\Omega (a_{ijkh} u_{k,h})_{,j} \dot{u}_i g\, dx$ and observing $g_{,i} = -\delta g x_i / r$, one may then establish

$$\begin{aligned} \frac{d}{dt}\frac{1}{2}\Big(\int_\Omega g\rho \dot{u}_i \dot{u}_i dx + \int_\Omega g a_{ijkh} u_{i,j} u_{k,h} dx \Big) \\ = \int_\Gamma g a_{ijkh} \dot{u}_i n_j u_{k,h} dS + \delta \int_\Omega g \frac{x_j}{r} a_{ijkh} u_{k,h} \dot{u}_i\, dx\,. \end{aligned} \tag{1.38}$$

In deriving (1.38) we have employed (1.36) and (1.37) on the boundary term which arises as $r \to \infty$. The term on Γ is zero since $u_i \equiv 0$ on Γ. Next, employ the bound (1.35) and the arithmetic - geometric mean inequality to derive from (1.38)

$$\begin{aligned}
&\frac{d}{dt}\frac{1}{2}\left(\int_\Omega g\rho \dot{u}_i\dot{u}_i dx + \int_\Omega g a_{ijkh} u_{i,j} u_{k,h} dx\right) \\
&\leq \delta M \int_\Gamma g|u_{k,h}|\,|\dot{u}_i|\,dx \\
&\leq \frac{\delta M}{2\alpha\rho_0}\int_\Omega g\rho \dot{u}_i\dot{u}_i dx + \frac{\delta M \alpha}{2a_0}\int_\Omega g a_{ijkh} u_{i,j} u_{k,h}\, dx
\end{aligned} \tag{1.39}$$

for $\alpha > 0$ to be selected. Pick now $\alpha = \sqrt{a_0/\rho_0}$ and define the weighted energy as

$$E(t) = \frac{1}{2}\left(\int_\Omega g\rho \dot{u}_i\dot{u}_i dx + \int_\Omega g a_{ijkh} u_{i,j} u_{k,h} dx\right). \tag{1.40}$$

Put $m = \delta M/2\rho_0\alpha$ and then (1.39) leads to

$$\frac{dE}{dt} \leq mE\,. \tag{1.41}$$

Integrate (1.41) from 0 to t to find

$$0 \leq E(t) \leq e^{mt}E(0) = 0,$$

where in the last step we note $\dot{u}_i \equiv 0$, $u_{i,j} \equiv 0$ when $t = 0$. Thus, $E \equiv 0$ on $[0,T)$ and so thanks to (1.18) $\nabla \mathbf{u} \equiv 0$ on $\Omega \times [0,T)$. Hence $u_i \equiv 0$ on Ω for all t. Therefore, we have established uniqueness of a solution to $\mathscr{P}$ when Ω is an unbounded exterior domain, even allowing for exponential solution growth at infinity.

1.3.6 Uniqueness Without Definiteness

Our aim in this section is to establish uniqueness for a solution to the boundary-initial value problem $\mathscr{P}$ of section (1.3.4) but by relaxing the conditions on the elastic coefficients a_{ijkh} and requiring no definiteness like (1.18) or (1.19). Instead we shall require a_{ijkh} to satisfy only the symmetry conditions (1.16). This uniqueness proof was originally given by Knops & Payne [140].

Before proceeding to the uniqueness question we require some preliminaries regarding convex functions. For $0 < \varepsilon < T$, a function $f(t)$ is convex on the interval $[\varepsilon, T]$ if

$$f'' \geq 0, \qquad t \in [\varepsilon, T]. \tag{1.42}$$

Then $f(t)$ lies below the straight line connecting $f(\varepsilon)$ and $f(T)$ and a simple geomtrical argument shows that

$$f(t) \leq \Big(\frac{t-\varepsilon}{T-\varepsilon}\Big) f(T) + \Big(\frac{T-t}{T-\varepsilon}\Big) f(\varepsilon). \tag{1.43}$$

Alternatively, one sees from (1.42) that $df'/dt \geq 0$ and so f' is an increasing function on (ε,t) and also on (t,T). Therefore, one may assert that

$$\int_\varepsilon^t f' ds \leq (t-\varepsilon) f'(t) \tag{1.44}$$

and

$$(T-t) f'(t) \leq \int_t^T f'(s) ds. \tag{1.45}$$

Carry out the integration in (1.44) and (1.45) and then one finds with a little rearrangement,

$$\frac{f(t)-f(\varepsilon)}{t-\varepsilon} \leq f'(t) \leq \frac{f(T)-f(t)}{T-t}.$$

Hence,

$$\frac{f(t)-f(\varepsilon)}{t-\varepsilon} \leq \frac{f(T)-f(t)}{T-t}.$$

If we solve this inequality for $f(t)$ then one arrives at (1.43).

In particular, when $f(t) = \log F(t)$ for some $F(t)$, inequality (1.43) becomes

$$\log F(t) \leq \left(\frac{t-\varepsilon}{T-\varepsilon}\right) \log F(T) + \left(\frac{T-t}{T-\varepsilon}\right) \log F(\varepsilon). \tag{1.46}$$

By taking the exponential, this leads to

$$F(t) \leq \big[F(T)\big]^{(t-\varepsilon)/(T-\varepsilon)} \big[F(\varepsilon)\big]^{(T-t)/(T-\varepsilon)}. \tag{1.47}$$

By letting $\varepsilon \to 0$ in (1.47) one finds

$$F(t) \leq \big[F(T)\big]^{t/T} \big[F(0)\big]^{(1-t/T)}. \tag{1.48}$$

To establish uniqueness for a solution to the boundary-initial value problem $\mathscr{P}$ of section 1.3.4 we let u_i be the difference solution which satisfies (1.30) with a_{ijkh} satisfying only the symmetry conditions (1.16). Then define a function $F(t)$ by

$$F(t) = <\rho u_i u_i>. \tag{1.49}$$

We calculate F' as

$$F'(t) = 2 < \rho u_i \dot{u}_i >, \tag{1.50}$$

and after a further differentiation we find

$$F''(t) = 2 < \rho \dot{u}_i \dot{u}_i > + 2 < \rho u_i \ddot{u}_i >. \tag{1.51}$$

The energy equation derived as (1.33) still holds and since $E(0) = 0$, this yields

$$< a_{ijkh}u_{i,j}u_{k,h} >= - < \rho \dot{u}_i \dot{u}_i > . \tag{1.52}$$

Substitute for $\rho \ddot{u}_i$ in (1.51) from $(1.30)_1$ and then with an integration by parts one shows from (1.51) that

$$F''(t) = 2 < \rho \dot{u}_i \dot{u}_i > - < a_{ijkh}u_{i,j}u_{k,h} >$$

and then an appeal to (1.52) allows one to deduce

$$F''(t) = 4 < \rho \dot{u}_i \dot{u}_i > . \tag{1.53}$$

From (1.49), (1.50) and (1.53) we now form $FF'' - (F')^2$ to see that

$$FF'' - (F')^2 = 4S^2, \tag{1.54}$$

where

$$S^2 =< \rho u_i u_i >< \rho \dot{u}_i \dot{u}_i > - < \rho u_i \dot{u}_i >^2 . \tag{1.55}$$

Note that $S^2 \geq 0$ by the Cauchy-Schwarz inequality.

Now proceed with a proof by contradiction. Assume u_i is not identically zero and for definiteness suppose $u_i \neq 0$ for $t > 0$ so that $F(t) > 0$ for $t \in [\varepsilon, T]$. Then we may divide (1.54) by F^2 on $[\varepsilon, T]$ to find

$$(\log F)'' \geq 0, \qquad t \in [\varepsilon, T].$$

Now fix $t \in (\varepsilon, T)$ and then from (1.46) we have

$$-\infty < \log F(t) \leq \left(\frac{t-\varepsilon}{T-\varepsilon}\right) \log F(T) + \left(\frac{T-t}{T-\varepsilon}\right) \log F(\varepsilon), \tag{1.56}$$

$t \in (\varepsilon, T)$. Since t is fixed $\log F(t)$ is fixed and $\log F(t) > -\infty$. Let $\varepsilon \to 0$ in (1.56) and the right hand side tends to $-\infty$ leading to a contradiction. We conclude that $F \equiv 0$ on $[0, T]$. Thus, the solution to $\mathscr{P}$ is unique.

1.3.7 An Alternative Proof

In this section we give an alternative proof of uniqueness for the problem studied in section 1.3.6. We still employ a logarithmic convexity method but now define $F(t)$ by

$$F(t) =< \rho u_i u_i > + \varepsilon, \tag{1.57}$$

for $\varepsilon > 0$ a constant.

By differentiation one may verify that expressions (1.50) and (1.53) still hold for F defined by (1.57). Thus, upon forming $FF'' - (F')^2$ one finds that

$$FF'' - (F')^2 = 4S^2 + 4\varepsilon < \rho \dot{u}_i \dot{u}_i >, \tag{1.58}$$

where S^2 is given by (1.55). Thus, again one finds for $t \in (0,T)$,

$$FF'' - (F')^2 \geq 0,$$

and so

$$(\log F)'' \geq 0, \qquad t \in [0,T].$$

Hence, $\log F$ is a convex function on the interval $t \in (0,T)$. In the present situation $F(0) = \varepsilon \neq 0$ and so we may utilize inequality (1.48). In this way, we deduce that with $F(t)$ given by (1.57), inequality (1.48) yields

$$< \rho u_i(t)u_i(t) > +\varepsilon \leq \left[< \rho u_i(T)u_i(T) > +\varepsilon\right]^{t/T} \varepsilon^{1-t/T}. \qquad (1.59)$$

Again, one proceeds by contradiction. Assume $u_i \neq 0$ on $\Omega \times (0,T)$ and then for fixed t, $< \rho u_i(t)u_i(t) > \neq 0$, and so we apply inequality (1.59). Let $\varepsilon \to 0$ and then (1.59) shows that

$$0 < K \leq < \rho u_i u_i > +\varepsilon \leq K_1^{t/T} \varepsilon^{(1-t/T)}, \qquad (1.60)$$

where $K_1 =< \rho u_i(T)u_i(T) > +\varepsilon$, and as $\varepsilon \to 0$ the right hand side of (1.60) tends to 0. Thus,

$$0 \leq < \rho u_i u_i > \leq 0$$

and so $u_i \equiv 0$ on $\Omega \times [0,T]$. Thus, uniqueness of a solution to $\mathscr{P}$ holds.

1.3.8 Kelvin - Voigt Theory

The basic equations of linear elastodynamics, equations (1.17) do not contain a dissipation term, i.e. there is no presence of $\dot{u}_i$ or its spatial derivatives. One may include such a term but the theory is then known as Kelvin - Voigt viscoelasticity, see Chirita *et al.* [40].

To derive a Kelvin-Voigt theory one must modify equation (1.14) and incorporate a dissipation term. For the anisotropic situation we may replace (1.14) by

$$\sigma_{ij} = a_{ijkh}e_{kh} + b_{ijkh}\dot{e}_{kh}. \qquad (1.61)$$

Here the elastic coefficients a_{ijkh} are required to satisfy the symmetry conditions (1.16), and we shall also require the coefficients b_{ijkh} to satisfy an analogous set of conditions, namely,

$$b_{ijkh} = b_{khij} = b_{jikh}. \qquad (1.62)$$

Thus, employing equation (1.61) in the momentum equation (1.12) one arrives at the governing equations for a Kelvin-Voigt elastic material

$$\rho \ddot{u}_i = (a_{ijkh}u_{k,h})_{,j} + (b_{ijkh}\dot{u}_{k,h})_{,j} + \rho f_i. \qquad (1.63)$$

For an isotropic material the Kelvin-Voigt stress law is, see Chirita *et al.* [40],

$$\sigma_{ij} = \lambda e_{mm}\delta_{ij} + 2\mu e_{ij} + \lambda^* \dot{e}_{mm}\delta_{ij} + 2\mu^* \dot{e}_{ij}\,, \tag{1.64}$$

where λ, μ are the Lamé coefficients, while λ^*, μ^* are coefficients of viscosity. This leads to the governing equations in the isotropic case as

$$\rho \ddot{u}_i = (\lambda + \mu)u_{j,ji} + \mu \Delta u_i + (\lambda^* + \mu^*)\dot{u}_{j,ji} + \mu^* \Delta \dot{u}_i + \rho f_i\,. \tag{1.65}$$

1.4 Definitions

We now give some definitions within the context of linear elastodynamics, although the concepts are easily extended to other theories. Hence, let $u_i(\mathbf{x},t)$ be a solution to the displacement boundary-initial value problem for equations (1.17) and suppose the symmetries (1.16) hold for the elastic coefficients. Thus, u_i solves the following boundary - initial value problem, $\mathscr{P}$,

$$\begin{aligned}
&\rho \ddot{u}_i = (a_{ijkh}u_{k,h})_{,j} + \rho f_i\,, \qquad \text{in } \Omega \times (0,T),\\
&u_i(\mathbf{x},t) = u_i^B(\mathbf{x},t), \qquad \text{on } \Gamma \times (0,T),\\
&u_i(\mathbf{x},0) = v_i(\mathbf{x}), \qquad \dot{u}_i(\mathbf{x},0) = w_i(\mathbf{x}), \qquad \mathbf{x} \in \Omega.
\end{aligned} \tag{1.66}$$

We say that a solution to $\mathscr{P}$ is *well-posed* in the sense of Hadamard, if the solution exists, if it is unique, and if it depends continuously on the data. In this section we restrict attention to continuous dependence upon the initial data, although later this could be extended to include boundary data, the coefficients themselves, and other data. This is known as structural stability, see chapters 5, 7. If a solution to $\mathscr{P}$ does not satisfy *any* one of these conditions we say the problem is *not well posed* or *improperly posed.*

We now give a simple example which illustrates that the concept of well posedness is not trivial.

1.4.1 Non-Uniqueness of Solution

Consider the one space dimensional equations of isotropic elasticity, equations (1.22). We suppose the body force is one which depends on the displacement u and we take $\rho f = u^\alpha$, for a constant α with $0 < \alpha < 1$. Here ρ, λ and μ are constants. Then suppose u satisfies the initial value problem

$$\begin{aligned}
&\rho \ddot{u} = (\lambda + 2\mu)u_{xx} + u^\alpha\,, \qquad x \in \mathbb{R},\ t > 0,\\
&u(x,0) = 0, \qquad \dot{u}(x,0) = 0, \qquad x \in \mathbb{R}.
\end{aligned}$$

Then we may immediately deduce that $u \equiv 0$ is one solution. However, one may also verify that there is another, non-zero, solution given by

$$u = Kt^{2/(1-\alpha)}, \qquad \text{where} \qquad K = \left[\frac{(1-\alpha)^2}{2\rho(1+\alpha)}\right]^{1/(1-\alpha)}.$$

This simple example shows that well posedness cannot be assumed. Even if a solution exists to a boundary-initial value problem, it may not be unique, or it may not depend continuously upon the initial data.

1.4.2 Stability

We now discuss the notion of stability for a solution to (1.66).

Let $\|\cdot\|$ and $|||\cdot|||$ be two norms on suitable spaces defined on the domain Ω. Let u_i^1 be a solution to (1.66) for data f_i, u_i^B, v_i^1 and w_i^1. Let u_i^2 be another solution to (1.66) for the same body force and boundary data, but for different initial data, say for data f_i, u_i^B, v_i^2 and w_i^2. Define u_i, v_i and w_i by

$$u_i = u_i^1 - u_i^2, \qquad v_i = v_i^1 - v_i^2, \qquad w_i = w_i^1 - w_i^2. \tag{1.67}$$

We say the solution u_i^1 is stable if given $\varepsilon > 0$, there exists $\delta > 0$, such that

$$\|\mathbf{u}(t)\| < \varepsilon \qquad \text{whenever} \qquad |||\mathbf{v},\mathbf{w}||| < \delta, \tag{1.68}$$

for all $t \in (0,T)$ where this holds for any u_i^2 which satisfies $(1.66)_1$ and $|||\mathbf{v},\mathbf{w}|||$ denotes a precise measure of the data functions $\mathbf{v}$ and $\mathbf{w}$. Another way to state (1.68) is

$$|||\mathbf{v},\mathbf{w}||| < \delta \quad \Longrightarrow \quad \|\mathbf{u}(t)\| < \varepsilon.$$

We say u_i^1 is asymptotically stable if it is stable, and in addition, $\|\mathbf{u}\| \to 0$ as $t \to \infty$. If $f_i = f_i(\mathbf{x})$ in (1.66) then u_i^1 will not, in general, be asymptotically stable, but this may happen if dissipation is present, cf. the situation for the Kelvin - Voigt equations, see exercise 1.11.

In this book we often make use of another kind of stability to that of (1.68). To introduce this we recollect that a real function f defined on an interval I is said to be Hölder continuous if there exists a constant α, $0 < \alpha \leq 1$, such that

$$|f(x) - f(y)| \leq M|x-y|^\alpha, \qquad \forall x, y \in I,$$

for some constant M. In the light of this we say a solution u_i^1 to (1.66) depends Hölder continuously on the initial data if

$$\|\mathbf{u}(t)\| < \|\mathbf{v}\|^\alpha, \tag{1.69}$$

for constants M, α with $0 < \alpha \leq 1$, where u_i and v_i are defined by (1.67).

Hölder continuous dependence on the initial data, or Hölder stability as it is sometimes known, may sometimes be recovered on a compact sub-interval of a time interval $[0,T)$, in an improperly posed problem when stability in the sense of (1.68) fails.

We have given a selected range of definitions of stability. One may extend this concept and there is an extensive range of definitions of various kinds of stability, see e.g. Hirsch & Smale [104], John [119], Knops & Wilkes [145].

1.4.3 Continuous Dependence upon Initial Data

In this section we prove continuous dependence on the initial data for a solution to (1.66) with the symmetries (1.16) but also when the elastic coefficients are positive definite in the sense that

$$a_{ijkh}\xi_{ij}\xi_{kh} \geq a_0\xi_{ij}\xi_{ij}, \qquad \forall \xi_{ij}. \tag{1.70}$$

Define u_i^1 and u_i^2 as in (1.67), and then from (1.66) we see that u_i satisfies the boundary - initial value problem

$$\begin{aligned} &\rho \ddot{u}_i = (a_{ijkh}u_{k,h})_{,j}, \qquad \text{in } \Omega\times(0,T),\\ &u_i(\mathbf{x},t) = 0, \qquad \text{on } \Gamma\times(0,T),\\ &u_i(\mathbf{x})0 = v_i(\mathbf{x}), \quad \dot{u}_i(\mathbf{x},0) = w_i(\mathbf{x}), \qquad \mathbf{x}\in\Omega. \end{aligned} \tag{1.71}$$

Multiply equation $(1.71)_1$ by $\dot{u}_i$ and integrate over Ω to find using $(1.71)_2$,

$$\frac{dE}{dt} = 0, \tag{1.72}$$

where

$$E(t) = \frac{1}{2} < \rho\dot{u}_i\dot{u}_i > + \frac{1}{2} < a_{ijkh}u_{i,j}u_{k,h} > . \tag{1.73}$$

Integrate (1.72) from 0 to t and then one obtains

$$E(t) = E(0) = \frac{1}{2} < \rho w_i w_i > + \frac{1}{2} < a_{ijkh}v_{i,j}v_{k,h} > . \tag{1.74}$$

Equation (1.74) already establishes stability in the sense of (1.68) if we let $E(t)$ be the $\|\cdot\|$ and $|||\cdot|||$ norms for u_i, v_i and w_i. However, if we employ (1.70) then from (1.74) we may also obtain

$$\frac{1}{2}a_0\|\nabla\mathbf{u}\|^2 \leq E(0) \tag{1.75}$$

which establishes stability as in (1.68) with the $|||\cdot|||$ norm on $\mathbf{v},\mathbf{w}$ being $E(0)$ whereas $\|\nabla\mathbf{u}\|$ represents the $\|\cdot\|$ norm, being the norm on $H_0^1(\Omega)$ since $(1.71)_2$ holds. Furthermore, if we employ Poincaré's inequality in (1.75) then we may see that

$$\|\mathbf{u}\|^2 \le \frac{2}{a_0\lambda_1} E(0), \tag{1.76}$$

where $\lambda_1 > 0$ is the first eigenvalue in the membrane problem for Ω. Inequality (1.76) yields another stability estimate of type (1.68) where the $|||\cdot|||$ norm is $E(0)$ and $\|\cdot\|$ is the norm on $L^2(\Omega)$.

In this section we have only considered stability in the displacement boundary - initial value problem where u_i is given on the boundary. One may consider stability with different boundary conditions such as prescribing the stress vector or other conditions, see e.g. Knops & Payne [139, 141, 144].

1.4.4 Continuous Dependence Without Definiteness

Current technology is requiring the creation of new materials. Many of these man made substances are likely to possess elastic coefficients which are not sign-definite. Among such classes of materials we may possibly include auxetic materials, see e.g. Greaves *et al.* [96], Sanami *et al.* [186]; graded systems for heat transport, see e.g. Jou *et al.* [128]; bodies with chiral structure, see e.g. Lakes [149], Ha *et al.* [100], Iesan & Quintanilla [116]; and certain composites, see e.g. Greaves *et al.* [96], Miller *et al.* [160]. From a mathematical viewpoint Poisson's ratio, cf. (1.21), may be negative, although many writers have argued that it should always be positive. Due to the creation of some new materials it is now a known fact that Poisson's ratio may be negative, see Xinchun & Lakes [228], with several other non typically elastic effects being reported in experiments, see e.g. Greaves *et al.* [96], Ha *et al.* [100]. Straughan [209] argues that in view of the range of new effects being discovered in elasticity it is desirable to analyse the case of indefinite elastic coefficients, especially in the context of a body with a multiple porosity structure. Straughan [209] also observes that a further motivation to require indefiniteness of the elastic coefficients arises from when the equations of elasticity are linearized about a state of nonlinear deformation, then the elastic tensor a_{ijkh} contains the effect of pre-stress. In that situation there is no compelling reason to demand that the elasticity tensor should be positive. This point is made on page 47 of the book by Straughan [203], and the pre-stress effects are also considered by Flavin & Green [85] and by Flavin [84].

In this section we investigate continuous dependence upon the initial data for a solution to (1.66) when a_{ijkh} are not required to be sign definite in any sense, and only the symmetry conditions (1.16) are required. We again let u_i^1 be a solution to (1.66) as in (1.67) with u_i^2 being another solution for different initial data, as in (1.67). Thus, the difference solution defined in (1.67) again satisfies the boundary - initial value problem (1.71). Due to the lack of definiteness of a_{ijkh} we employ a logarithmic convexity method. This was first demonstrated by Knops & Payne [138], where more detailed results may be found.

Let us observe that the energy equation (1.74) again holds with $E(t)$ given by (1.73).

The proof of continuous dependence is split into two parts.

1.4.5 Non-Positive Initial Energy

Here we suppose $E(0) \leq 0$. Define $F(t)$ by

$$F(t) =< \rho u_i(\mathbf{x},t)u_i(\mathbf{x},t) > . \tag{1.77}$$

Then, by differentiation

$$F'(t) = 2 < \rho u_i\dot{u}_i >, \tag{1.78}$$

and

$$F''(t) = 2 < \rho\dot{u}_i\dot{u}_i > +2 < \rho u_i\ddot{u}_i > .$$

Substitute for $\rho\ddot{u}_i$ from $(1.71)_1$ and integrate by parts on the a_{ijkh} term using the boundary conditions $(1.71)_2$ to find

$$F'' = 2 < \rho\dot{u}_i\dot{u}_i > -2 < a_{ijkh}u_{i,j}u_{k,h} > .$$

Note that from the energy equation (1.74) one has

$$-2 < a_{ijkh}u_{i,j}u_{k,h} >= 2 < \rho\dot{u}_i\dot{u}_i > -4E(0). \tag{1.79}$$

Next, substitute for the a_{ijkh} term from (1.79) to arrive at

$$F'' = 4 < \rho\dot{u}_i\dot{u}_i > -4E(0). \tag{1.80}$$

Now form $FF'' - (F')^2$ using (1.77), (1.78) and (1.80), to find

$$FF'' - (F')^2 = 4S^2 - 4FE(0), \tag{1.81}$$

where

$$S^2 =< \rho u_iu_i >< \rho\dot{u}_i\dot{u}_i > - < \rho u_i\dot{u}_i >^2, \tag{1.82}$$

and $S^2 \geq 0$ by the Cauchy-Schwarz inequality. Thus, since $E(0) \leq 0$ from (1.81) we divide by F^2 to find for $t \in (0,T)$,

$$(\log F)'' \geq 0, \qquad t \in (0,T). \tag{1.83}$$

Hence, F satisfies inequality (1.48) and so

$$F(t) \leq \big[F(T)\big]^{t/T}\big[F(0)\big]^{(1-t/T)}.$$

Suppose now the solution u_i belongs to the class where $F(T)$ is bounded in the sense that $F(T) \leq M < \infty$ for a constant M. Then one finds

$$F(t) \leq M^{t/T}\big[F(0)\big]^{(1-t/T)}, \tag{1.84}$$

for $t \in [0,T)$. Thus, the solution u_i^1 to (1.66) depends Hölder continuously upon the initial data in the sense of (1.69) on compact sub-intervals of $[0,T)$, where $\alpha = 1-t/T < 1$. The norm used in (1.69) in this case is the weighted $L^2(\Omega)$ norm given by (1.77).

1.4.6 Positive Initial Energy

In this section $E(0) > 0$. While equation (1.80) still holds one cannot deduce (1.83) since $E(0) > 0$. Hence, one needs to work with a function different from F, see Knops & Payne [138]. Define $G(t)$ by

$$G(t) = \log\left[F(t) + 2E(0)\right] + t^2, \tag{1.85}$$

where $F(t)$ is given by (1.77).

By calculation one finds

$$G' = \frac{F'}{F(t)+2E(0)} + 2t, \tag{1.86}$$

and

$$G'' = \frac{F''}{F(t)+2E(0)} - \frac{(F')^2}{[F(t)+2E(0)]^2} + 2. \tag{1.87}$$

One forms the combination

$$[F(t)+2E(0)]^2G'' = [F(t)+2E(0)]F'' - (F')^2 + 2[F(t)+2E(0)]^2 \tag{1.88}$$

and the idea is to deduce $G'' \geq 0$, so G is a convex function of t.

Now, use (1.78) and (1.80) in the right hand side of (1.88). We see that

$$\begin{aligned}
[F(t)+2E(0)]^2G'' =& [F(t)+2E(0)]\left[4 < \rho \dot{u}_i\dot{u}_i > -4E(0)\right] \\
& -4 < \rho u_i\dot{u}_i >^2 + 2[F(t)+2E(0)]^2 \\
=& 4S^2 + 2F^2 + 4E(0)F + 8E(0) < \rho \dot{u}_i\dot{u}_i > ,
\end{aligned} \tag{1.89}$$

where $S^2 \geq 0$ is given by (1.82). The right hand side of (1.89) is non-negative and hence $G'' \geq 0$ and so G is a convex function of t.

Hence, G satisfies an inequality like (1.46) but with $\varepsilon = 0$. Therefore,

$$G(t) \leq \frac{t}{T}G(T) + \left(1 - \frac{t}{T}\right)G(0). \tag{1.90}$$

Thus, employing (1.85)

$$\log[F(t)+2E(0)] \leq \frac{t}{T}\log[F(T)+2E(0)]+Tt-t^2 \\ +\left(1-\frac{t}{T}\right)\log[F(0)+2E(0)]. \tag{1.91}$$

After taking the exponential of this inequality one sees that

$$F(t)+2E(0) \leq \exp\left[t(T-t)\right] \\ \times \left[F(T)+2E(0)\right]^{t/T}\left[F(0)+2E(0)\right]^{(1-t/T)}. \tag{1.92}$$

Thus, for the class of solutions satisfying the bound $F(T) \leq M < \infty$ inequality (1.92) leads to a Hölder continuous dependence upon the initial data estimate of the form (1.69). The estimate is thus of form

$$F(t)+2E(0) \leq K[F(0)+2E(0)]^{\alpha},$$

for t in a compact sub-interval of $[0,T)$, where $0 < \alpha = 1-t/T < 1$, and $K = \exp\left[t(T-t)\right][M+2E(0)]^{t/T}$.

1.5 Overview

Chapter 1 has introduced basic concepts in linear elasticity and some of the techniques to be employed throughout. In chapter 2 we review the development of multi-porosity in elasticity and present relevant equations. Chapter 3 reviews the development of another description of double porosity elasticity by a voids distribution approach. In chapter 4 we provide a comparison of the double porosity and double voids theories of chapters 2 and 3 by using various mathematical techniques. Chapter 5 concentrates on deriving uniqueness and stability results in multi-porosity theories. Chapter 6 shows how one may establish uniqueness in a variety of multi-porosity elasticity or thermoelasticity theories without requiring sign-definiteness of the elastic coefficients. In chapter 7 we study continuous dependence upon the initial data and structural stability questions in multi-porosity elasticity.

Chapters 8 and 9 analyse wave motion in double porosity elasticity from both the viewpoints of chapters 2 and 3 and we develop a nonlinear acceleration wave theory for both classes of material.

In the final chapter 10, we introduce new material which investigates the effect of temperature waves, also known as second sound, in a double porosity elastic body.

1.6 Exercises

Exercise 1.1. Show that the strong ellipticity condition (1.24) is weaker than the condition of positive - definiteness (1.23), i.e. show (1.23) implies (1.24).

Exercise 1.2. In two-dimensions there are 16 elastic coefficients a_{ijkh}, $i,j,k,h = 1,2$. Show that the Cauchy symmetries (1.15) reduce the number of independent elastic coefficients to 9.

Exercise 1.3. In three-dimensions show that for Cauchy elasticity the symmetries (1.15) reduce the 81 independent elastic coefficients a_{ijkh} down to 36. Show further that if the Green symmetry (1.16) is further imposed then there are 21 independent coefficients.

Exercise 1.4. MacCullagh's theory of light is based on the equations

$$\rho \ddot{u}_i = t_{ij,j} \tag{1.93}$$

where the stress tensor, t_{ij}, is given in the isotropic case by

$$t_{ij} = K w_{ij}$$

where

$$w_{ij} = \frac{1}{2}(u_{i,j} - u_{j,i}),$$

see Truesdell & Toupin [221], pp. 727–729.

Consider the boundary-initial value problem $\mathscr{P}$ consisting of (1.93) defined on Ω with either

a) u_i given on Γ,

b) $\operatorname{curl}\mathbf{u}$ given on Γ,

or

c) u_i given on Γ_1, $\operatorname{curl}\mathbf{u}$ given on Γ_2, $\Gamma_1 \cup \Gamma_2 = \Gamma$. Show that the solution to $\mathscr{P}$ is unique.

Hint. Show that equations (1.93) may be written as

$$\rho \ddot{u}_i = -\frac{K}{2}(\operatorname{curl}\operatorname{curl}\mathbf{u})_i .$$

Multiply this equation by $\dot{u}_i$ and integrate over Ω to obtain

$$E(t) = E(0),$$

where

$$E(t) = <\rho \dot{u}_i \dot{u}_i> + \frac{K}{2}\|\operatorname{curl}\mathbf{u}\|^2.$$

Exercise 1.5. Let $\mathbf{u} = (u,v,w)$. Write the equations of isotropic elasticity (1.22) in full in terms of the components (u,v,w).

Exercise 1.6. Let $\mathbf{u} = (u,v,w)$. Write equations (1.17) in full in terms of the components (u,v,w), for the cubic system (1.25), (1.26).

Exercise 1.7. Let $\mathbf{u} = (u,v,w)$. Write equations (1.17) in full in terms of the components (u,v,w), for the transversely isotropic system (1.27).

Exercise 1.8. Let u_i be a function in Ω such that $u_i = 0$ on Γ. Denote the symmetric and skew-symmetric parts of $u_{i,j}$ by e_{ij} and s_{ij}, i.e.

$$e_{ij} = \frac{1}{2}(u_{i,j} + u_{j,i}), \qquad s_{ij} = \frac{1}{2}(u_{i,j} - u_{j,i}).$$

Show that

$$\int_\Omega e_{ij}e_{ij}dx - \int_\Omega s_{ij}s_{ij}dx = \int_\Omega (u_{i,i})^2 dx. \tag{1.94}$$

Denote by V the potential energy of an isotropic linear elastic body, so that

$$V = \frac{1}{2}\mu \int_\Omega u_{i,j}u_{i,j}dx + \frac{1}{2}(\lambda + \mu) \int_\Omega u_{i,j}u_{j,i}dx,$$

cf. equations (1.22). By using (1.94) and integrating by parts show that V may be written in the alternative form, cf. Borchardt [23], Thomson [218],

$$V = \frac{1}{2}(\lambda + 2\mu) \int_\Omega (u_{i,i})^2 dx + \mu \int_\Omega s_{ij}s_{ij}dx. \tag{1.95}$$

Exercise 1.9. Consider the displacement boundary - initial value problem for equations (1.22). Show that the solution is unique provided the coefficients λ and μ satisfy the strongly semi - elliptic conditions

$$\mu \geq 0, \quad \lambda + 2\mu \geq 0, \tag{1.96}$$

cf. (1.24).
Hint. Let u_i^1 and u_i^2 be solutions to (1.22) for the same boundary data u_i^B, same initial data $u_i(\mathbf{x},0) = v_i(\mathbf{x})$, $\dot{u}_i(\mathbf{x},0) = w_i(\mathbf{x})$, and for the same body force f_i. Define $u_i = u_i^1 - u_i^2$ and then use an energy method to show that

$$\frac{1}{2} < \rho\dot{u}_i\dot{u}_i > = -\frac{\mu}{2} < u_{i,j}u_{i,j} > - \frac{(\lambda + \mu)}{2} < u_{i,j}u_{j,i} > .$$

Use expressions (1.94) and (1.95) to deduce

$$\frac{1}{2} < \rho\dot{u}_i\dot{u}_i > = -\frac{(\lambda + 2\mu)}{2} \int_\Omega (u_{i,i})^2 dx - \mu \int_\Omega s_{ij}s_{ij}dx.$$

Use the strongly semi-elliptic conditions (1.96) to deduce $\dot{u}_i \equiv 0$ and then $u_i \equiv 0$ and so uniqueness.

Exercise 1.10. Suppose a_{ijkh} and b_{ijkh} are non-negative in the Kelvin-Voigt equations (1.63). Consider the boundary-initial value problem, $\mathscr{P}$, consisting of equations (1.63) defined on $\Omega \times \{t > 0\}$ together with u_i given on Γ and $u_i, \dot{u}_i$ defined when $t = 0$. By using a standard energy technique show that the solution to $\mathscr{P}$ is unique.

Exercise 1.11. Consider the same boundary-initial value problem as that of exercise 1.10, but now do not require any definiteness of a_{ijkh}, although b_{ijkh} are still non-negative. Show that the solution to $\mathscr{P}$ is unique.
Hint. Use a logarithmic convexity argument with the function

$$F =< \rho u_i u_i > + \int_0^t < b_{ijkh} u_{i,j} u_{k,h} > ds.$$

You should find

$$F' = 2 < \rho u_i \dot{u}_i > +2 \int_0^t < b_{ijkh} u_{i,j} \dot{u}_{k,h} > ds,$$
$$F'' = 2 < \rho \dot{u}_i \dot{u}_i > -2 < a_{ijkh} u_{i,j} u_{k,h} >,$$

and employ the energy equation

$$E(t) + \int_0^t < b_{ijkh} \dot{u}_{i,j} \dot{u}_{k,h} > ds = E(0),$$

where

$$E(t) = \frac{1}{2} < \rho \dot{u}_i \dot{u}_i > + \frac{1}{2} < a_{ijkh} u_{i,j} u_{k,h} > .$$

Exercise 1.12. Consider the boundary-initial value problem, $\mathscr{P}_1$, for the Kelvin-Voigt equations (1.63) on $\Omega \times \{t > 0\}$, with $f_i = 0$, with $u_i = 0$ on $\Gamma \times \{t > 0\}$ and with $u_i, \dot{u}_i$ given at $t = 0$. Suppose

$$b_{ijkh} \xi_{ij} \xi_{kh} \geq b_0 \xi_{ij} \xi_{ij}, \qquad \forall\, \xi_{ij}, \tag{1.97}$$

together with

$$\rho_m \geq \rho(\mathbf{x}) \geq \rho_0, \qquad \forall\, \mathbf{x} \in \Omega, \tag{1.98}$$

and

$$a_{ijkh} \xi_{ij} \xi_{kh} \geq \frac{b_0 \lambda_1}{2\rho_m} b_{ijkh} \xi_{ij} \xi_{kh}, \qquad \forall\, \xi_{ij}, \tag{1.99}$$

where $b_0 > 0, \rho_m > 0, \rho_0 > 0$ are constants and λ_1 is the constant in the Poincaré inequality for Ω. Show that a solution to $\mathscr{P}_1$ will decay exponentially under the conditions above.
Hint. Transform equations (1.63) into equations for a new variable v_i given by $v_i = u_i e^{\lambda t}$, where $b_0 \lambda_1 / 2\rho_m > \lambda > 0$. Multiply the resulting equation for v_i by $\dot{v}_i$ and integrate over Ω to find

$$\frac{dL}{dt} + < b_{ijkh} \dot{v}_{i,j} \dot{v}_{k,h} > -2\lambda < \rho \dot{v}_i \dot{v}_i >= 0$$

where

$$L(t) = \frac{1}{2} < \rho \dot{v}_i \dot{v}_i > + \frac{\lambda^2}{2} < \rho v_i v_i > + \frac{1}{2} < a_{ijkh} v_{i,j} v_{k,h} > - \frac{\lambda}{2} < b_{ijkh} v_{i,j} v_{k,h} > .$$

Use Poincaré's inequality together with (1.97) and (1.98) to deduce

$$\frac{dL}{dt} \leq 0.$$

Integrate this from 0 to t and return to the u_i variables. Use (1.99) and then deduce

$$< \rho u_i u_i > \leq \frac{2}{\lambda^2} L(0) e^{-2\lambda t}.$$

Exercise 1.13. Consider the continuous dependence upon the initial data question for the Kelvin-Voigt equations (1.63) when the elastic coeffcients satisfy only the symmetry conditions (1.16) but are not sign definite. Suppose the coefficients b_{ijkh} satisfy the symmetry conditions (1.62) and are also non-negative, i.e.

$$b_{ijkh}\xi_{ij}\xi_{kh} \geq 0, \qquad \forall\, \xi_{ij} \neq 0.$$

Suppose u_i satisfy the boundary and initial conditions

$$u_i(\mathbf{x},t) = u_i^B(\mathbf{x},t) \qquad \text{on } \Gamma \times (0,T), \tag{1.100}$$

and

$$u_i(\mathbf{x},0) = v_i(\mathbf{x}), \qquad \dot{u}_i(\mathbf{x},0) = w_i(\mathbf{x}), \quad \mathbf{x} \in \Omega. \tag{1.101}$$

Define the energy function $E(t)$ by

$$E(t) = \frac{1}{2} < \rho \dot{u}_i \dot{u}_i > + \frac{1}{2} < a_{ijkh} u_{i,j} u_{k,h} > .$$

Show that the solution to (1.63), (1.100) and (1.101) depends continuously upon the initial data in the Hölder sense, on compact sub-intervals of $(0,T)$, when $E(0) \leq 0$. *Hint.* Use a logarithmic convexity argument with the function

$$F(t) =< \rho u_i u_i > + \int_0^t < b_{ijkh} u_{i,j} u_{k,h} > ds + (T-t) < b_{ijkh} v_{i,j} v_{k,h} >$$

where $0 < t < T$. Show

$$F' = 2 < \rho u_i \dot{u}_i > +2 \int_0^t < b_{ijkh} u_{i,j} \dot{u}_{k,h} > ds$$

and

$$F'' = 4 < \rho \dot{u}_i \dot{u}_i > +4 \int_0^t < b_{ijkh} \dot{u}_{i,j} \dot{u}_{k,h} > ds - 4E(0).$$

Then deduce $FF'' - (F')^2 \geq 0$ and invoke inequality (1.48).

Exercise 1.14. Consider the boundary-initial value problem for equations (1.17) with the elastic coefficients obeying the symmetries (1.16). Thus, u_i satisfies the boundary-initial value problem

$$
\begin{aligned}
&\rho \ddot{u}_i = (a_{ijkh}u_{k,h})_{,j}\,, && \text{in } \Omega \times \{t > 0\}, \\
&u_i(\mathbf{x},t) = 0, && \text{on} \times \{t > 0\}, \\
&u_i(\mathbf{x},0) = v_i(\mathbf{x}), && \dot{u}_i(\mathbf{x},0) = w_i(\mathbf{x}),
\end{aligned}
\tag{1.102}
$$

for v_i and w_i prescribed functions.

Derive the energy equation

$$E(t) = E(0)$$

where

$$E(t) = \frac{1}{2} < \rho \dot{u}_i \dot{u}_i > + \frac{1}{2} < a_{ijkh}u_{i,j}u_{k,h} > .$$

A) When $E(0) < 0$, show that $< \rho u_i u_i >$ grows exponentially in time.

B) When $E(0) = 0$ and $< \rho v_i w_i > > 0$ show also that $< \rho u_i u_i >$ grows exponentially in time.

Hint.

For A) show that one may select β and τ so that

$$F(t) =< \rho u_i u_i > + \beta(t+\tau)^2 \tag{1.103}$$

is a logarithmically convex function. Show by integrating the convexity inequality $(\log F)'' \geq 0$ one has

$$F(t) \geq F(0) \exp\left(\frac{F'(0)t}{F(0)}\right). \tag{1.104}$$

The choice $\beta = -2E(0)$ works and then one selects τ to ensure $F'(0) > 0$.

For B) take F as in (1.103) with $\beta = 0$. Show that F is logarithmically convex and then use (1.104) and the given restriction on v_i and w_i.

For further details of growth results in equations encompassing those of linear elasticity see Knops & Payne [142].

Chapter 2
Models for Double and Triple Porosity

In this chapter we review several of the mathematical models which have appeared in the literature and are capable of describing the evolutionary behaviour of a double or triple porosity elastic medium, in some appropriate sense.

There are many different ways to approach the problem of modelling a double or triple porosity elastic body and it is not our objective to attempt a comprehensive review. Many of the early approaches are discussed in detail in e.g. Chen [36] or in Hornung & Showalter [106], and in the references therein. We describe models which are compatible with a modern continuum mechanics approach and where possible we employ arguments from continuum thermodynamics.

Choo *et al.* [45] point out that "*several studies have adopted the double-porosity framework for coupled solid deformation and fluid flow in porous media*", but they also draw attention to the fact that "*there has been significant disagreement in the theoretical formulation of the problem*".

As indicated, there are many different approaches to a theoretical formulation of double or triple porosity elasticity theory. We review several which have led to much success with stability, wave and uniqueness studies. It is worth pointing out that we do not deal with homogenization techniques although there is a successful homogenization approach to double porosity elasticity, cf. e.g. Arbogast *et al.* [6], Chen [35], Rohan *et al.* [179], Showalter & Visarraga [195], and the references therein. Additionally, internal variables have been employed in the description of a double porosity model, see e.g. Konyukhov & Pankratov [146], and constitutive theory has been developed for specific geological materials, see e.g. Cariou *et al.* [29]. Very recent research has extended the double porosity theory from a saturated porous elastic body to one where the doubly porous elastic body is not totally saturated, see e.g. Choo *et al.* [45]. This is clearly an important development which will have a multitude of interesting applications. One application area for double porosity elasticity with a partially saturated porous medium which immediately springs to mind is landslides, see e.g. Borja *et al.* [24], Borja & White [25], Montrasio *et al.* [162], Scotto di Santolo & Evangelista [187].

B. Straughan, *Mathematical Aspects of Multi–Porosity Continua*, Advances in Mechanics and Mathematics 38, https://doi.org/10.1007/978-3-319-70172-1_2

2.1 Quasi-Equilibrium Model

Throughout this chapter we shall be mainly concerned with an elastic body with double porosity. The pressure in the macro pores will be denoted by $p(\mathbf{x},t)$ whereas that in the micro pores will be denoted by $q(\mathbf{x},t)$. The quasi-equilibrium theory we describe is due to Berryman & Wang [14] but before we come to this we consider the development of the equations for the pressures themselves, cf. Chen [36], Hornung & Showalter [106], and the references therein.

According to Chen [36], Barenblatt & Zheltov [12] introduced a radically different approach to modelling a naturally fractured reservoir and initiated a theory of flow through a double porosity body. Chen [36] oberserves that the pressures p and q satisfy the equations

$$\begin{aligned} c_1\phi_1\frac{\partial p}{\partial t} &= \frac{k_1}{\tilde{\mu}}\Delta p - \frac{\alpha}{\tilde{\mu}}(p-q), \\ c_2\phi_2\frac{\partial q}{\partial t} &= \frac{k_2}{\tilde{\mu}}\Delta q + \frac{\alpha}{\tilde{\mu}}(p-q), \end{aligned} \tag{2.1}$$

where $c_\alpha, \phi_\alpha, k_\alpha$, $\alpha = 1,2$, are, respectively, the isothermal compressibility, porosity, and the permeability in the macro and micro pores. The coefficient α is to be determined for particular materials and $\tilde{\mu}$ is the dynamic viscosity of the fluid. Chen [36] notes that Barenblatt *et al.* [13] simplified equations (2.1) and assumed the flow through the matrix block system and the storage capacity in the micro system may be neglected. This leads to the simplified set of equations

$$\begin{aligned} c_1\phi_1\frac{\partial p}{\partial t} &= -\frac{\alpha}{\tilde{\mu}}(p-q), \\ \frac{k_2}{\tilde{\mu}}\Delta q + \frac{\alpha}{\tilde{\mu}}(p-q) &= 0. \end{aligned} \tag{2.2}$$

Chen [36] observes that one may eliminate either p or q from (2.2) to see that either of these functions satisfies the partial differential equation

$$\frac{\partial u}{\partial t} = \frac{k_2}{\tilde{\mu}\phi_1 c_1}\Delta u + \frac{k_2}{\alpha}\Delta\frac{\partial u}{\partial t}, \tag{2.3}$$

where $u = p$ or q.

Chen [36] further proceeds to explain that another simplified model which may arise from (2.1) is suggested by Warren & Root [225]. This system has form

$$\begin{aligned} c_1\phi_1\frac{\partial p}{\partial t} &= -\frac{\alpha}{\tilde{\mu}}(p-q), \\ c_2\phi_2\frac{\partial q}{\partial t} &= \frac{k_2}{\tilde{\mu}}\Delta q + \frac{\alpha}{\tilde{\mu}}(p-q). \end{aligned} \tag{2.4}$$

Berryman & Wang [14] effectively extended equations (2.1) and introduced the elastic displacement of the porous skeleton into the system. Let us denote by $u_i(\mathbf{x},t)$ the elastic displacement. Berryman & Wang [14] begin by writing the equations for an elastic body with a single porosity and they say this is from the low frequency limit of the theory of Biot, cf. Biot [18]. For a single porosity Berryman & Wang [14] write equations for the displacement, u_i, and pressure, p, as

$$
\begin{aligned}
&\frac{3K}{2(1+\nu)}u_{j,ji}+\frac{3(1-2\nu)K}{2(1+\nu)}\Delta u_i=\alpha p_{,i}\,,\\
&\frac{\alpha}{BK_u}\frac{\partial p}{\partial t}+\alpha\frac{\partial u_{j,j}}{\partial t}=\frac{k}{\tilde{\mu}}\Delta p.
\end{aligned}
\tag{2.5}
$$

In these equations K is the drained bulk modulus of the elastic body, ν is Poisson's ratio, K_u is the undrained bulk modulus, B is a parameter described in Berryman & Wang [14] and $\alpha=(1-K/K_u)B$. Again, k and $\tilde{\mu}$ are the permeability and dynamic viscosity of the fluid. Showalter [194] develops a rigorous existence, regularity and uniqueness theory for an isotropic, quasi-equilibrium single porosity model like (2.5).

Let us note that equation $(2.5)_1$ is the balance of momentum equation with the acceleration term $\rho\ddot{u}_i$ omitted, ρ being the density of the elastic body. This gives rise to the so-called "quasi-equilibrium" theory.

Berryman & Wang [14] generalize equations (2.5) together with equations (2.1) to write a quasi-equilibrium model for a linear elastic body with double porosity as

$$
\begin{aligned}
&\frac{3K}{2(1+\nu)}u_{j,ji}+\frac{3(1-2\nu)K}{2(1+\nu)}\Delta u_i=\alpha_1 p_{,i}+\alpha_2 q_{,i}\,,\\
&\frac{\alpha_1}{B_1K_u^1}\dot{p}+\alpha_1\dot{u}_{j,j}=\frac{k_1}{\tilde{\mu}}\Delta p-\gamma(p-q)\,,\\
&\frac{\alpha_2}{B_2K_u^2}\dot{q}+\alpha_2\dot{u}_{j,j}=\frac{k_2}{\tilde{\mu}}\Delta q+\gamma(p-q)\,,
\end{aligned}
\tag{2.6}
$$

where a superposed dot denotes $\partial/\partial t$, e.g. $\dot{p}=\partial p/\partial t$. In addition, $\alpha_1,\alpha_2,B_1,B_2$ and γ are coefficients to be determined experimentally. We mention that Showalter & Walkington [196] and Cook and Showalter [55] provide a rigorous investigation of well posedness of double porosity models with microstructure.

2.2 Berryman-Wang Model

We have described a quasi-equilibrium model for a double porosity elastic material in section 2.1. Even though this model was presented by Berryman & Wang [14] we retain the nomenclature Berryman & Wang model for that presented in this section.

The theory of Berryman & Wang [15] retains the inertia in the momentum equation for the elastic skeleton, i.e. the $\rho\ddot{u}_i$ term. However, they also introduce the fluid displacements in the macro and micro pores, respectively, as $v_i(\mathbf{x},t)$ and $w_i(\mathbf{x},t)$. They derive their system of equations for $(\mathbf{u},\mathbf{v},\mathbf{w})$ as

$$\begin{aligned}
&\rho_{11}\ddot{u}_i+\rho_{12}\ddot{v}_i+\rho_{13}\ddot{w}_i+(b_{12}+b_{13})\dot{u}_i-b_{12}\dot{v}_i-b_{13}\dot{w}_i\\
&\qquad=\Big(K_u+\frac{\mu}{3}\Big)u_{j,ji}+\mu\Delta u_i+K_u(B_1\zeta^1_{,i}+B_2\zeta^2_{,i}),\\
&\rho_{12}\ddot{u}_i+\rho_{22}\ddot{v}_i+\rho_{23}\ddot{w}_i-b_{12}\dot{u}_i+(b_{12}+b_{23})\dot{v}_i-b_{23}\dot{w}_i=-(1-v_2)\phi^1p_{,i}\,,\\
&\rho_{13}\ddot{u}_i+\rho_{23}\ddot{v}_i+\rho_{33}\ddot{w}_i-b_{13}\dot{u}_i-b_{23}\dot{v}_i+(b_{13}+b_{23})\dot{w}_i=-v_2\phi^2q_{,i}\,,
\end{aligned}\tag{2.7}$$

where v_2 is the total volume fraction of the micro porosity, ζ_1,ζ_2 are the flux volume accumulations per unit bulk volume in the macro and micro pores, and the other coefficients appearing in (2.7) are constants. Berryman & Wang [15] go into great detail about how one actually calculates these coefficients in a real situation and they give details for Berea sandstone with water. The fluid increments ζ_1 and ζ_2 are defined by

$$\zeta_1=-(1-v_2)\phi^1(v_{i,i}-u_{i,i}),\qquad \zeta_2=-v_2\phi^2(w_{i,i}-u_{i,i}),\tag{2.8}$$

and Berryman & Wang [15] show that for suitable coefficients k_{ij} these are related to the pressures by the equations

$$\begin{aligned}
\dot{\zeta}_1&=k_{11}\Delta p+k_{12}\Delta q,\\
\dot{\zeta}_2&=k_{21}\Delta p+k_{22}\Delta q.
\end{aligned}\tag{2.9}$$

Berryman & Wang [15] also employ the relation

$$\begin{pmatrix}-u_{i,i}\\ \zeta_1\\ \zeta_2\end{pmatrix}=\begin{pmatrix}a_{11}&a_{12}&a_{13}\\ a_{12}&a_{22}&a_{23}\\ a_{13}&a_{23}&a_{33}\end{pmatrix}\begin{pmatrix}p_c\\ p\\ q\end{pmatrix}\tag{2.10}$$

where a_{ij} may be calculated and p_c is the external confining pressure. Denote by $A=a^{-1}$ the inverse of the matrix $a\equiv(a_{ij})$ in (2.10). Then rewrite (2.10) in terms of A as

$$\begin{pmatrix}p_c\\ p\\ q\end{pmatrix}=A\begin{pmatrix}-u_{i,i}\\ \zeta_1\\ \zeta_2\end{pmatrix}\tag{2.11}$$

If we now define $\mathscr{A}$ to be the 2×2 matrix

$$\mathscr{A}=\begin{pmatrix}A_{22}&A_{23}\\ A_{23}&A_{33}\end{pmatrix}$$

then we see from the last two components of equation (2.11) that

$$\mathscr{A}\begin{pmatrix}\zeta_1\\ \zeta_2\end{pmatrix}=\begin{pmatrix}p+A_{12}u_{i,i}\\ q+A_{13}u_{i,i}\end{pmatrix}$$

and then

$$\begin{pmatrix}\zeta_1\\ \zeta_2\end{pmatrix} = \frac{1}{|\mathscr{A}|}\begin{pmatrix}A_{33} & -A_{23}\\ -A_{23} & A_{22}\end{pmatrix}\begin{pmatrix}p+A_{12}u_{i,i}\\ q+A_{13}u_{i,i}\end{pmatrix},$$

where $|\mathscr{A}| = \det\mathscr{A}$. Thus,

$$\begin{aligned}\zeta_1 &= \frac{1}{|\mathscr{A}|}\left\{A_{33}p - A_{23}q + (A_{33}A_{12} - A_{23}A_{13})u_{i,i}\right\},\\ \zeta_2 &= \frac{1}{|\mathscr{A}|}\left\{A_{22}q - A_{23}p + (A_{22}A_{13} - A_{12}A_{23})u_{i,i}\right\}.\end{aligned} \tag{2.12}$$

If we rewrite equations (2.7) and (2.9) eliminating ζ_1 and ζ_2 with the aid of (2.12) we obtain a system of equations in u_i, v_i, w_i, p and q where the p and q equations have some resemblance to equations (2.6) of section 2.1.

We return to the equations of this section in chapter 8 where Rayleigh waves and Love waves are discussed in section 8.1, cf. Dai *et al.* [59], and Dai & Kuang [58].

2.3 Svanadze Model

In a series of papers Svanadze and his co-workers, see Svanadze [211, 212, 214, 213], Ciarletta *et al.* [51], Scarpetta & Svanadze [189], Scarpetta *et al.* [190], Svanadze & Scalia [216, 217], have derived linear equations for an isotropic elastic body with double porosity or with triple porosity, incorporating also temperature, i.e. equations for a double or triple porosity thermoelastic body. We review these developments here although we also allow the body to be anisotropic, features which for applications may well be very relevant, cf. Zhao & Chen [230].

We begin with a double porosity elastic material at constant temperature. Svanadze [212, 214] begins with the equation of balance of linear momentum for the elastic solid of form

$$\rho\ddot{u}_i = t_{ji,j}, \tag{2.13}$$

where $\rho(\mathbf{x})$ is the density of the elastic skeleton and t_{ij} is the (symmetric) Cauchy stress tensor. Svanadze [212] employs a constitutive equation for the stress which is linear in the strain $e_{ij} = (u_{i,j} + u_{j,i})/2$ and in the pressures p and q. For an anisotropic material this relation may be written as

$$t_{ij} = t_{ij}^e - \beta_{ij}p - \gamma_{ij}q, \tag{2.14}$$

where t_{ij}^e is the elastic part of the stress, namely

$$t_{ij}^e = a_{ijkh}e_{kh}, \tag{2.15}$$

and β_{ij} and γ_{ij} are symmetric tensors which may depend on $\mathbf{x}$. In addition, there are two equations governing the behaviour of the pressures p and q. These are

$$v^1_{i,i} + \beta_{ij}\dot{e}_{ij} + \dot{\zeta}_1 + \gamma(p-q) = 0, \tag{2.16}$$

and

$$v^2_{i,i} + \gamma_{ij}\dot{e}_{ij} + \dot{\zeta}_2 - \gamma(p-q) = 0, \tag{2.17}$$

where $\mathbf{v}^1$ and $\mathbf{v}^2$ are the fluid velocities in the macro and micro pores, respectively, while ζ_1 and ζ_2 represent the increment of fluid in the macro pores and micro pores, respectively.

To motivate equations (2.16) and (2.17) we consider an arbitrary volume V in the double porosity body. The boundary of V we denote by ∂V. Then we may write conservation laws for the macro phase and the micro phase as

$$\oint_{\partial V} (v^1_i n_i + \beta_{ij}\dot{u}_j n_i)dS = -\int_V \{\dot{\zeta}_1 + \gamma(p-q)\}dV \tag{2.18}$$

and

$$\oint_{\partial V} (v^2_i n_i + \gamma_{ij}\dot{u}_j n_i)dS = -\int_V \{\dot{\zeta}_2 + \gamma(q-p)\}dV\,. \tag{2.19}$$

The increment of fluid functions ζ_1 and ζ_2 will depend only on the pressures p and q. Hence, from a physical viewpoint, equations (2.18) and (2.19) show that if the velocities are such that the boundary particles are moving outward the pressure inside V decreases and vice versa. The γ terms are due to interactions between the fluids in the macro and micro pores. We believe that physically equations (2.18) and (2.19) make sense. To proceed from equations (2.18) and (2.19) we employ the divergence theorem on the boundary terms and then we use the symmetry of β_{ij} and γ_{ij} to obtain

$$\int_V \{v^1_{i,i} + \beta_{ij}\dot{e}_{ij} + \dot{\zeta}_1 + \gamma(p-q)\}dV = 0, \tag{2.20}$$

and

$$\int_V \{v^2_{i,i} + \gamma_{ij}\dot{e}_{ij} + \dot{\zeta}_2 + \gamma(q-p)\}dV = 0. \tag{2.21}$$

Let now the integrands in equations (2.20) and (2.21) be denoted by I_1 and I_2. Thus,

$$\int_V I_\alpha dV = 0, \qquad \alpha = 1,2.$$

Suppose now $I_\alpha(\mathbf{x}) \neq 0$ for some $\mathbf{x} \in V$. For definiteness, suppose $I_1 > 0$ at $\mathbf{x}$. Then there is a neighbourhood V_1 around $\mathbf{x}$ such that $I_1 > 0$ in V_1. Thus

$$\int_V I_1 dV > 0.$$

The volume V is arbitrary and so we take $V = V_1$. This yields a contradiction and so $I_1 \equiv 0$ and equation (2.16) follows. A similar argument leads to equation (2.17).

Svanadze [212] argues that

$$\zeta_1 = \alpha p, \qquad \zeta_2 = \beta q, \tag{2.22}$$

whereas Svanadze [214] suggests

$$\zeta_1 = \alpha p + \alpha_1 q, \qquad \zeta_2 = \alpha_1 p + \beta q. \tag{2.23}$$

We allow the coefficients α, β and α_1 to depend on $\mathbf{x}$. Svanadze [214] invokes arguments of Khalili [133] which suggest that if one neglects cross coupling via the α_1 terms then incorrect predictions of flow and deformation may result, see also Berryman & Wang [15] and Masters *et al.* [159]. Svanadze [212, 214] employs a form of Darcy's law to relate $\mathbf{v}^1$ to ∇p and $\mathbf{v}^2$ to ∇q and for an anisotropic medium we may write

$$v_i^1 = -k_{ij} p_{,j}, \qquad v_i^2 = -m_{ij} q_{,j}. \tag{2.24}$$

If one now employs equations (2.24) in (2.16) and (2.17) and one employs (2.14) and (2.15) in (2.13) then together with equation (2.22) one derives the system of equations

$$\rho \ddot{u}_i = \frac{\partial}{\partial x_j}\Big(a_{ijkh}\frac{\partial u_k}{\partial x_h}\Big) - \frac{\partial}{\partial x_j}\Big(\beta_{ij} p\Big) - \frac{\partial}{\partial x_j}\Big(\gamma_{ij} q\Big) + \rho f_i, \tag{2.25}$$

together with

$$\alpha \dot{p} = \frac{\partial}{\partial x_i}\Big(k_{ij}\frac{\partial p}{\partial x_j}\Big) - \gamma(p-q) - \beta_{ij}\frac{\partial \dot{u}_i}{\partial x_j} + \rho s_1(\mathbf{x},t), \tag{2.26}$$

and

$$\beta \dot{q} = \frac{\partial}{\partial x_i}\Big(m_{ij}\frac{\partial q}{\partial x_j}\Big) + \gamma(p-q) - \gamma_{ij}\frac{\partial \dot{u}_i}{\partial x_j} + \rho s_2(\mathbf{x},t). \tag{2.27}$$

Observe that in equations (2.25) - (2.27) we have included the body force f_i and the supply terms s_1 and s_2. These are prescribed quantities we may control.

If we instead employ the consitutive theory (2.23) then we arrive at the system of equations

$$\rho \ddot{u}_i = \frac{\partial}{\partial x_j}\Big(a_{ijkh}\frac{\partial u_k}{\partial x_h}\Big) - \frac{\partial}{\partial x_j}\Big(\beta_{ij} p\Big) - \frac{\partial}{\partial x_j}\Big(\gamma_{ij} q\Big) + \rho f_i, \tag{2.28}$$

and

$$\alpha_1 \dot{p} + \alpha_2 \dot{q} = \frac{\partial}{\partial x_i}\Big(k_{ij}\frac{\partial p}{\partial x_j}\Big) - \gamma(p-q) - \beta_{ij}\frac{\partial \dot{u}_i}{\partial x_j} + \rho s_1(\mathbf{x},t), \tag{2.29}$$

together with

$$\beta_1 \dot{p} + \beta_2 \dot{q} = \frac{\partial}{\partial x_i}\Big(m_{ij}\frac{\partial q}{\partial x_j}\Big) + \gamma(p-q) - \gamma_{ij}\frac{\partial \dot{u}_i}{\partial x_j} + \rho s_2(\mathbf{x},t). \tag{2.30}$$

To use equations (2.25) - (2.27) or (2.28) - (2.30) we must incorporate them into a boundary value problem. To do this we suppose equations (2.25) - (2.27) or (2.28) - (2.30) hold on the region $\Omega \times (0,T)$ where Ω is a domain in $\mathbb{R}^3$ and $(0,T)$ is a time interval, $0 < T < \infty$. For the applications studied in this book the domain Ω will have a boundary and this is denoted by Γ. The region Γ will be assumed smooth enough to allow application of the divergence theorem. Boundary conditions must be prescribed and we consider the following ones on the displacement and the pressures,

$$u_i(\mathbf{x},t) = h_i(\mathbf{x},t), \quad p(\mathbf{x},t) = p^B(\mathbf{x},t), \quad q(\mathbf{x},t) = q^B(\mathbf{x},t), \tag{2.31}$$

on $\Gamma \times (0,T)$. The functions h_i, p^B and q^B are prescribed. One must also prescribe initial conditions on the elastic displacement, the elastic velocity, and on the pressures, and we consider those of form

$$\begin{aligned} u_i(\mathbf{x},0) &= u_i^0(\mathbf{x}), & \dot{u}_i(\mathbf{x},0) &= v_i^0(\mathbf{x}), \\ p(\mathbf{x},0) &= p^0(\mathbf{x}), & q(\mathbf{x},0) &= q^0(\mathbf{x}). \end{aligned} \tag{2.32}$$

In (2.32) $\mathbf{x} \in \Omega$, and u_i^0, v_i^0, p^0 and q^0 are prescribed data functions. The boundary - initial value problem composed of equations (2.25) - (2.27) or (2.28) - (2.30) together with the boundary conditions (2.31) and the initial conditions (2.32) will be denoted by $\mathscr{P}$.

2.3.1 Thermoelastic Model

Section 2.3 restricts attention to the isothermal situation for anisotropic double porosity elasticity. In practical applications one encounters many situations where deformation may be induced by changing thermal effects. This is especially true in the case of elastic bodies with double or triple porosities where thermal stresses may introduce micro cracking, see e.g. David *et al.* [61], Gelet *et al.* [89], Homand-Etienne & Houpert [105], Kim & Hosseini [137]. Recognizing the importance of thermal effects Svanadze [214] and Gelet *et al.* [89] together with Scarpetta *et al.* [190] and Scarpetta & Svanadze [189] have developed equations to describe the evolutionary behaviour of a linear, isotropic double porosity thermoelastic material. Straughan [208] generalizes the theory of Gelet *et al.* [89] and Svanadze [214] and gives a system of partial differential equations which are appropriate to study the behaviour in time of an *anisotropic* double porosity elastic body. To present these equations we introduce the temperature field $\theta(\mathbf{x},t)$.The system of equations given by Straughan [208] may be written as

$$\begin{aligned} &\rho \ddot{u}_i = (a_{ijkh}u_{k,h})_{,j} - (\beta_{ij}p)_{,j} - (\gamma_{ij}q)_{,j} - (a_{ij}\theta)_{,j} + \rho f_i, \\ &\alpha_1 \dot{p} + \alpha_2 \dot{q} + \gamma_1 \dot{\theta} = (k_{ij}p_{,j})_{,i} - \gamma(p-q) - \beta_{ij}\dot{u}_{i,j} + \rho s_1\,, \\ &\beta_1 \dot{p} + \beta_2 \dot{q} + \gamma_2 \dot{\theta} = (m_{ij}q_{,j})_{,i} + \gamma(p-q) - \gamma_{ij}\dot{u}_{i,j} + \rho s_2, \\ &a\dot{\theta} + a_{ij}\dot{u}_{i,j} + \gamma_1 \dot{p} + \gamma_2 \dot{q} = (\kappa_{ij}\theta_{,j})_{,i} + \rho r(\mathbf{x},t)\,. \end{aligned} \tag{2.33}$$

In equations (2.33) $\kappa(\mathbf{x})$ is the thermal diffusivity, $a(\mathbf{x})$ is the thermal inertia coefficient, γ_1 and γ_2 are other inertia coefficients, $a_{ij}(\mathbf{x})$ are thermal coupling coefficients, and r is a prescribed heat supply. For clarity we observe that equations (2.33) constitute a system of six partial differential equations in the six variables u_1, u_2, u_3, p, q and θ.

The system of partial differential equations (2.33) is a thermoelastic equivalent of equations (2.28) - (2.30). For resolution of some problems it may be sufficient to work with equations which incorporate thermal effects but are a generalization of system (2.25) - (2.27). In this case an appropriate system of equations is, cf. Straughan [208],

$$\begin{aligned} \rho\ddot{u}_i &= (a_{ijkh}u_{k,h})_{,j} - (\beta_{ij}p)_{,j} - (\gamma_{ij}q)_{,j} - (a_{ij}\theta)_{,j} + \rho f_i\,, \\ \alpha\dot{p} &= (k_{ij}p_{,j})_{,i} - \gamma(p-q) - \beta_{ij}\dot{u}_{i,j} + \rho s_1\,, \\ \beta\dot{q} &= (m_{ij}q_{,j})_{,i} + \gamma(p-q) - \gamma_{ij}\dot{u}_{i,j} + \rho s_2\,, \\ a\dot{\theta} + a_{ij}\dot{u}_{i,j} &= (\kappa_{ij}\theta_{,j})_{,i} + \rho r. \end{aligned} \tag{2.34}$$

2.3.2 Svanadze Isotropic Models

Svanadze [212, 214] actually derives models for isotropic elastic bodies with multiple porosities. For completeness we record the isotropic versions of equations (2.25) - (2.27) and (2.28) - (2.30) below. Corresponding to (2.25) - (2.27) one has

$$\rho\ddot{u}_i = \mu\Delta u_i + (\lambda+\mu)\frac{\partial}{\partial x_i}\Big(\frac{\partial u_j}{\partial x_j}\Big) - \hat{\beta}\frac{\partial p}{\partial x_i} - \hat{\gamma}\frac{\partial q}{\partial x_i} + \rho f_i\,, \tag{2.35}$$

together with

$$\alpha\dot{p} = k\Delta p - \gamma(p-q) - \hat{\beta}\frac{\partial \dot{u}_i}{\partial x_i} + \rho s_1(\mathbf{x},t)\,, \tag{2.36}$$

and

$$\beta\dot{q} = m\Delta q + \gamma(p-q) - \hat{\gamma}\frac{\partial \dot{u}_i}{\partial x_i} + \rho s_2(\mathbf{x},t)\,. \tag{2.37}$$

The analogous isotropic equations to (2.28) - (2.30) are

$$\rho\ddot{u}_i = \mu\Delta u_i + (\lambda+\mu)\frac{\partial}{\partial x_i}\Big(\frac{\partial u_j}{\partial x_j}\Big) - \hat{\beta}\frac{\partial p}{\partial x_i} - \hat{\gamma}\frac{\partial q}{\partial x_i} + \rho f_i\,, \tag{2.38}$$

with

$$\alpha_1\dot{p} + \alpha_2\dot{q} = k\Delta p - \gamma(p-q) - \hat{\beta}\frac{\partial \dot{u}_i}{\partial x_i} + \rho s_1(\mathbf{x},t)\,, \tag{2.39}$$

and additionally

$$\beta_1\dot{p} + \beta_2\dot{q} = m\Delta q + \gamma(p-q) - \hat{\gamma}\frac{\partial \dot{u}_i}{\partial x_i} + \rho s_2(\mathbf{x},t)\,. \tag{2.40}$$

The isotropic linear thermoelasticity equations analogous to (2.33) are

$$\begin{aligned}
&\rho\ddot{u}_i = \mu\Delta u_i + (\lambda+\mu)u_{j,ji} - \hat{\beta}p_{,i} - \hat{\gamma}q_{,i} - \hat{a}\theta_{,i} + \rho f_i\,,\\
&\alpha_1\dot{p} + \alpha_2\dot{q} + \gamma_1\dot{\theta} = k\Delta p - \gamma(p-q) - \hat{\beta}\dot{u}_{i,i} + \rho s_1\,,\\
&\beta_1\dot{p} + \beta_2\dot{q} + \gamma_2\dot{\theta} = m\Delta q + \gamma(p-q) - \hat{\gamma}\dot{u}_{i,i} + \rho s_2\,,\\
&a\dot{\theta} + \hat{a}\dot{u}_{i,i} + \gamma_1\dot{p} + \gamma_2\dot{q} = \hat{k}\Delta\theta + \rho r.
\end{aligned} \tag{2.41}$$

The set of equations represented by (2.41) is a system of six partial differential equations in the six variables u_1, u_2, u_3, p, q and θ.

The linear isotropic double porosity thermoelastic system analogous to (2.34) where the cross "inertia" coefficients are omitted is

$$\begin{aligned}
&\rho\ddot{u}_i = \mu\Delta u_i + (\lambda+\mu)u_{j,ji} - \hat{\beta}p_{,i} - \hat{\gamma}q_{,i} - \hat{a}\theta_{,i} + \rho f_i\,,\\
&\alpha\dot{p} = k\Delta p - \gamma(p-q) - \hat{\beta}\dot{u}_{i,i} + \rho s_1\,,\\
&\beta\dot{q} = m\Delta q + \gamma(p-q) - \hat{\gamma}\dot{u}_{i,i} + \rho s_2\,,\\
&a\dot{\theta} + \hat{a}\dot{u}_{i,i} = \hat{k}\Delta\theta + \rho r.
\end{aligned} \tag{2.42}$$

2.3.3 Triple Porosity Elastic Model

Svanadze [215] develops a set of equations for an isotropic linear elastic body which allows for a triple porosity structure. In this case the body has the usual macro porosity and one associates with this a fluid pressure in the pores which is denoted by $p(\mathbf{x},t)$. There is further a structure of pores on a smaller scale to which is attached a porosity known as meso porosity. Associated to the meso porosity one attaches a pressure of the fluid which we denote by $q(\mathbf{x},t)$. However, there is a further even smaller porosity structure which gives rise to a micro porosity. The pressure of the fluid in the micro pores is denoted by $s(\mathbf{x},t)$.

Straughan [208] generalizes the approach of Svanadze [215] and describes a system of linear partial differential equations for an anisotropic elastic body with a triple porosity internal structure. Denoting again by $u_i(\mathbf{x},t)$ the elastic displacement, the equations given in Straughan [208] are

$$\begin{aligned}
&\rho\ddot{u}_i = (a_{ijkh}u_{k,h})_{,j} - (\beta_{ij}p)_{,j} - (\gamma_{ij}q)_{,j} - (\omega_{ij}s)_{,j} + \rho f_i\,,\\
&\alpha_1\dot{p} + \alpha_2\dot{q} + \alpha_3\dot{s} = (k_{ij}p_{,j})_{,i} + \lambda_{12}(p-q) + \lambda_{13}(p-s) - \beta_{ij}\dot{u}_{i,j} + \rho s_1\,,\\
&\beta_1\dot{p} + \beta_2\dot{q} + \beta_3\dot{s} = (m_{ij}q_{,j})_{,i} + \lambda_{21}(q-p) + \lambda_{23}(q-s) - \gamma_{ij}\dot{u}_{i,j} + \rho s_2\,,\\
&\gamma_1\dot{p} + \gamma_2\dot{q} + \gamma_3\dot{s} = (\ell_{ij}s_{,j})_{,i} + \lambda_{31}(s-p) + \lambda_{32}(s-q) - \omega_{ij}\dot{u}_{i,j} + \rho s_3\,,
\end{aligned} \tag{2.43}$$

where f_i is a prescribed body force and s_1, s_2, s_3 are externally given supply functions. All of the coefficients in equations (2.43) may depend on the spatial variable $\mathbf{x}$.

The incorporation of pressure terms in (2.43) follows Svanadze [215] who allows interaction between the pressures in the macro, meso and micro phases. There is an interaction between the pressures in the macro and meso pores which is represented by the terms involving λ_{12} and λ_{21}. The interaction between the pressures in the meso and micro pores is given by the terms involving λ_{23} and λ_{32}. Finally, (2.43) allows for interaction between the pressures in the macro and micro pores via the terms involving λ_{13} and λ_{31}.

Equations (2.43) incorporate cross inertia effects via the $\alpha_2, \alpha_3, \beta_1, \beta_3, \gamma_1$ and γ_2 terms. They also account for interactions between the macro pore pressure and the micro pore pressure. It is likely that in some problems one may ignore these effects as argued in e.g. Bai *et al.* [9, 8], Bai & Roegiers [10], Kuznetsov & Nield [148], Straughan [206]. For such a scenario one may be able to replace equations (2.43) by a reduced system of form

$$\begin{aligned}
\rho\ddot{u}_i &= (a_{ijkh}u_{k,h})_{,j} - (\beta_{ij}p)_{,j} - (\gamma_{ij}q)_{,j} - (\omega_{ij}s)_{,j} + \rho f_i\,,\\
\alpha\dot{p} &= (k_{ij}p_{,j})_{,i} - \gamma(p-q) - \beta_{ij}\dot{u}_{i,j} + \rho s_1\,,\\
\beta\dot{q} &= (m_{ij}q_{,j})_{,i} + \gamma(p-q) + \xi(s-q) - \gamma_{ij}\dot{u}_{i,j} + \rho s_2\,,\\
\varepsilon\dot{s} &= (\ell_{ij}s_{,j})_{,i} - \xi(s-q) - \omega_{ij}\dot{u}_{i,j} + \rho s_3\,.
\end{aligned} \tag{2.44}$$

Notice that equations (2.44) still account for interactions between macro - meso and meso - micro phases.

While Svanadze [215] includes all pressure interaction terms as in equations (2.43) it is worth recording the system of equations which arises from (2.44) when the body is isotropic. In this case the system of partial differential equations governing the evolutionary behaviour of a linear isotropic elastic body with triple porosity may be written as

$$\begin{aligned}
\rho\ddot{u}_i &= \lambda\Delta u_i + (\lambda+\mu)u_{j,ji} - \beta p_{,i} - \hat{\gamma}q_{,i} - \omega s_{,i} + \rho f_i\,,\\
\alpha\dot{p} &= k\Delta p - \gamma(p-q) - \beta\dot{u}_{i,i} + \rho s_1\,,\\
\delta\dot{q} &= m\Delta q + \gamma(p-q) + \xi(s-q) - \hat{\gamma}\dot{u}_{i,i} + \rho s_2\,,\\
\varepsilon\dot{s} &= \ell\Delta s - \xi(s-q) - \omega\dot{u}_{i,i} + \rho s_3\,.
\end{aligned} \tag{2.45}$$

In equations (2.45) the coefficients ρ, λ, μ together with $\beta, \hat{\gamma}, \omega, \alpha, \delta, \varepsilon, \gamma$ and ξ are constants and Δ is the Laplace operator, i.e. $\Delta \equiv \partial^2/\partial x_i \partial x_i$.

2.3.4 Triple Porosity Thermoelasticity

As has been observed in section 2.3.1 thermal stresses may well be responsible for producing cracking in multi-porosity elastic materials. Straughan [209] studied well posedness for a system of partial differential equations which is effectively a combination of equations (2.33) for double porosity thermoelasticity and equations (2.43) for isothermal triple porosity elasticity. The equations written by Straughan

[209] which account for cross inertia effects and interactions between the macro and meso pressures, the meso and micro pressures, and the macro and micro pressures, are

$$\begin{aligned}
&\rho\ddot{u}_i = (a_{ijkh}u_{k,h})_{,j} - (\beta_{ij}p)_{,j} - (\gamma_{ij}q)_{,j} - (\omega_{ij}s)_{,j} - (a_{ij}\theta)_{,j} + \rho f_i\,,\\
&\alpha\dot{p} + \alpha_1\dot{q} + \alpha_2\dot{s} + \alpha_3\dot{\theta} = (k_{ij}p_{,j})_{,i} - \beta_{ij}\dot{u}_{i,j} - \gamma(p-q) - \omega(p-s) + \rho s_1\,,\\
&\alpha_1\dot{p} + \beta\dot{q} + \beta_1\dot{s} + \beta_2\dot{\theta} = (m_{ij}q_{,j})_{,i} - \gamma_{ij}\dot{u}_{i,j} + \gamma(p-q) - \xi(q-s) + \rho s_2\,,\\
&\alpha_2\dot{p} + \beta_1\dot{q} + \varepsilon\dot{s} + \varepsilon_1\dot{\theta} = (\ell_{ij}q_{,j})_{,i} - \omega_{ij}\dot{u}_{i,j} + \xi(q-s) + \omega(p-s) + \rho s_3\,,\\
&\alpha_3\dot{p} + \beta_2\dot{q} + \varepsilon_1\dot{s} + a\dot{\theta} = (r_{ij}\theta_{,j})_{,i} - a_{ij}\dot{u}_{i,j} + \rho r\,.
\end{aligned} \tag{2.46}$$

All the coefficients in (2.46) are allowed to depend on $\mathbf{x}$, $\rho(\mathbf{x}) > 0$ is the density, and $a_{ijkh}(\mathbf{x})$ are the elastic coefficients. The tensors $\beta_{ij}, \gamma_{ij}, \omega_{ij}, a_{ij}, k_{ij}, m_{ij}, \ell_{ij}$ and r_{ij} are all symmetric, and the body force f_i, heat supply r, and source terms s_1, s_2 and s_3 are all prescribed functions.

For some applications it may be adequate to neglect the cross inertia effects and the interactions between the pressures in the macro and micro pores. In this case one would arrive at a system of equations of form

$$\begin{aligned}
&\rho\ddot{u}_i = (a_{ijkh}u_{k,h})_{,j} - (\beta_{ij}p)_{,j} - (\gamma_{ij}q)_{,j} - (\omega_{ij}s)_{,j} - (a_{ij}\theta)_{,j} + \rho f_i\,,\\
&\alpha\dot{p} = (k_{ij}p_{,j})_{,i} - \beta_{ij}\dot{u}_{i,j} - \gamma(p-q) + \rho s_1\,,\\
&\beta\dot{q} = (m_{ij}q_{,j})_{,i} - \gamma_{ij}\dot{u}_{i,j} + \gamma(p-q) - \xi(q-s) + \rho s_2\,,\\
&\varepsilon\dot{s} = (\ell_{ij}q_{,j})_{,i} - \omega_{ij}\dot{u}_{i,j} + \xi(q-s) + \rho s_3\,,\\
&a\dot{\theta} = (r_{ij}\theta_{,j})_{,i} - a_{ij}\dot{u}_{i,j} + \rho r\,.
\end{aligned} \tag{2.47}$$

In the case where the elastic body is isotropic, equations (2.47) will reduce to the forms

$$\begin{aligned}
&\rho\ddot{u}_i = \lambda\Delta u_i + (\lambda+\mu)u_{j,ji} - \beta p_{,i} - \hat{\gamma}q_{,i} - \omega s_{,i} - \hat{a}\theta_{,i} + \rho f_i\,,\\
&\alpha\dot{p} = k\Delta p - \gamma(p-q) - \beta\dot{u}_{i,i} + \rho s_1\,,\\
&\delta\dot{q} = m\Delta q + \gamma(p-q) + \xi(s-q) - \hat{\gamma}\dot{u}_{i,i} + \rho s_2\,,\\
&\varepsilon\dot{s} = \ell\Delta s - \xi(s-q) - \omega\dot{u}_{i,i} + \rho s_3\,,\\
&a\dot{\theta} = r\Delta\theta - \hat{a}\,\dot{u}_{i,i} + \rho r.
\end{aligned} \tag{2.48}$$

2.4 Khalili-Selvadurai Model

Khalili & Selvadurai [134] present a more general model than those seen so far. They again have a momentum equation for u_i and equations governing the pressures p and q. However, they treat a non-isothermal situation in which the elastic solid may have a temperature $T_s(\mathbf{x},t)$ whereas the temperature of the fluid in the macro pores is $T_1(\mathbf{x},t)$ while the temperature of the fluid in the micro pores is $T_2(\mathbf{x},t)$. In general T_s, T_1 and T_2 are different. This is thus a double porosity version of a local thermal non-equilibrium situation, cf. Straughan [206]. Khalili & Selvadurai [134] produce equations also for T_s, T_1 and T_2. Their system is thus one of eight interconnected equations governing the behaviour of the eight variables u_1, u_2, u_3, p, q, T_s, T_1 and T_2.

The system of equations of Khalili & Selvadurai [134] is conveniently given in equations (27a)-(27b) of their paper.

We complete this section by referring to the very interesting contribution of Choo *et al.* [45]. These writers produce a model to describe the evolutionary behaviour of a double porosity elastic body when the fluid does *not* saturate the whole of the pores, i.e. an unsaturated medium. They develop a theory of infinitesimal deformation but they pointedly remark that to the best of their knowledge, "*this is the first time that a poromechanical framework for double porosity media has been developed from direct application of the principles of thermodynamics.*"

2.5 Exercises

Exercise 2.1. Consider the solution to equations (2.1) on a bounded domain Ω, with $p = q = 0$ on the boundary Γ. Show that $p, q \to 0$ in the $L^2(\Omega)$ norm.

Exercise 2.2. Consider the solution to equations (2.2) on a bounded domain Ω, with $p = q = 0$ on the boundary Γ. Show that $p, q \to 0$ in a suitable measure.

Exercise 2.3. Consider the solution to equations (2.4) on a bounded domain Ω, with $p = q = 0$ on the boundary Γ. Show that $p, q \to 0$ in a suitable measure.

Exercise 2.4. Equations (2.46) and (2.47) are appropriate for a description of the behaviour of a linear elastic body with a triple porosity structure. Suggest systems of equations analogous to (2.46) and (2.47) for the description of an isothermal elastic body with a quadruple porosity structure.

Exercise 2.5. Equations (2.46) and (2.47) are appropriate for a description of the behaviour of a linear elastic body with a triple porosity structure. Suggest systems of equations analogous to (2.46) and (2.47) for the description of a thermoelastic body with a quadruple porosity structure.

Exercise 2.6. Equations (2.46) and (2.47) are appropriate for a description of the behaviour of a linear elastic body with a triple porosity structure. Suggest systems of equations analogous to (2.46) and (2.47) for the description of an isothermal elastic body with a quintuple porosity structure.

Exercise 2.7. Equations (2.46) and (2.47) are appropriate for a description of the behaviour of a linear elastic body with a triple porosity structure. Suggest systems of equations analogous to (2.46) and (2.47) for the description of a thermoelastic body with a quintuple porosity structure.

Exercise 2.8. In equations (2.25) - (2.27) the connectivity terms between the macro and micro pore structures are those involving $\gamma(p-q)$. Suggest a set of partial differential equations for an elastic body with double porosity in which the connectivities are not linear.

Chapter 3
Double Porosity and Voids

3.1 Basics

Our aim in this chapter is to describe a theory of nonlinear thermoelasticity where the body contains a double porosity system of voids. There is a void distribution associated to the macro pores, but additionally there is another void distribution due to the micro pores. This work was developed by Iesan & Quintanilla [115].

Rather than begin immediately with the double porosity voids theory we firstly introduce the classical theory of nonlinear thermoelasticity with no voids present. We then describe the theory of nonlinear thermoelasticity with a single distribution of voids. The theory of elastic materials containing voids was developed by Nunziato & Cowin [165]. A theory of nonlinear thermoelasticity with voids was developed by Iesan [109] and this is also described in detail in the book by Iesan [111]. The theory of elastic materials containing voids is particularly useful to describe nonlinear wave motion and accounts well for the elastic behaviour of the matrix, being a generalisation of nonlinear elasticity theory. Interestingly, while there are many studies involving the linearised theory of elastic materials with voids, see e.g. Ciarletta & Iesan [50] or Iesan [111], analysis of the fully nonlinear equations is seen less frequently, see e.g. Iesan [109, 112, 113], Ciarletta & Straughan [52], Ciarletta *et al.* [53], Straughan [202], chapter 7. It is worth observing that Iesan [114] has developed a fully nonlinear theory of a thermoelastic body containing voids which accounts also for highly viscoelastic effects. This theory of thermoviscoelastic materials with voids is developed for a fully nonlinear theory using continuum thermodynamic arguments and we observe that the linearized version of the theory derived by Iesan [114] bears resemblance to a generalization of the Kelvin-Voigt theory of elasticity, cf. section 1.3.8.

The basic idea of including voids in a continuous body is due to Goodman & Cowin [94], although they developed constitutive theory appropriate to a fluid. This they claim is more appropriate to flow of a granular medium. Acceleration waves in the Goodman-Cowin theory of granular media were studied by Nunziato & Walsh [166, 167]. For a reader interested in the theory of voids I would suggest first reading

B. Straughan, *Mathematical Aspects of Multi–Porosity Continua*, Advances in Mechanics and Mathematics 38, https://doi.org/10.1007/978-3-319-70172-1_3

the article of Goodman & Cowin [94], and then progressing to the theory of elastic materials with voids as given by Nunziato & Cowin [165]. General descriptions of the theory of elastic materials with voids and various applications are given in the books of Ciarletta & Iesan [50] and Iesan [111]. Continuous dependence on the coupling coefficients of the voids theory (a structural stability problem) is studied by Chirita *et al.* [39].

We now introduce basic notation and then we progress to the classical theory of nonlinear thermoelasticity.

3.1.1 Bodies and Their Configurations

We consider a body B deformed from a *reference configuration* at time $t = 0$ to a *current configuration* at time t.

Points in the reference configuration are labelled by boldface notation $\mathbf{X}$ or indicial notation X_A. In the current configuration $\mathbf{X} \to \mathbf{x}$. The mapping is thus

$$\mathbf{x} = \mathbf{x}(\mathbf{X}, t) \tag{3.1}$$

or

$$x_i = x_i(X_A, t). \tag{3.2}$$

The coordinates X_A are *material* (or *Lagrangian*) coordinates whereas x_i are *spatial coordinates* (*Eulerian coordinates*).

In elasticity we need the displacement vector $\mathbf{u}$ of a typical particle from $\mathbf{X}$ in the reference configuration to $\mathbf{x}$ at time t, so

$$u_i(X_A, t) = x_i(X_A, t) - X_i. \tag{3.3}$$

The velocity of a particle v_i is

$$v_i(X_A, t) = \frac{\partial x_i}{\partial t}\Big|_{\mathbf{X}\ \text{constant}}.$$

In fluid mechanics we usually use the inverse of (3.1) to write $v_i = v_i(x_j, t)$ - this is the spatial description, i.e. that following the particle.

3.1.2 The Deformation Gradient Tensor

The deformation gradient tensor F_{iA} is defined by

$$F_{iA} = \frac{\partial x_i}{\partial X_A}. \tag{3.4}$$

From (3.3) we find the displacement gradient as

$$u_{i,A} = \frac{\partial u_i}{\partial X_A} = \frac{\partial x_i}{\partial X_A} - \delta_{iA} = F_{iA} - \delta_{iA}\,. \tag{3.5}$$

3.1.3 Conservation of Mass

The relation

$$\rho_0 = \rho \det \mathbf{F} \tag{3.6}$$

is the conservation of mass in Lagrangian form. (Recall that in Eulerian form this is

$$\frac{\partial \rho}{\partial t} + (\rho v_i)_{,i} = 0.)$$

N.B. If the material is incompressible then $\rho =$ constant so $\rho_0/\rho = 1$. Therefore, in an incompressible material the deformation must satisfy

$$\det \mathbf{F} = 1.$$

(See Spencer [199], pp. 91–95.)

3.2 The Equations of Nonlinear Thermoelasticity

Key to the theory is the tensor π_{Ai} which is called the *Piola - Kirchoff* stress tensor (useful in elasticity because it refers back to the reference configuration). The equation of linear momentum for a thermoelastic body is

$$\rho_0 \ddot{x}_i = \frac{\partial \pi_{Ai}}{\partial X_A} + \rho_0 f_i, \tag{3.7}$$

or

$$\rho_0 \frac{\partial^2 x_i}{\partial t^2}\bigg|_{\mathbf{X}} = \frac{\partial \pi_{Ai}}{\partial X_A} + \rho_0 f_i, \tag{3.8}$$

see Spencer [199], eq. (9.38). In equation (3.7) the term f_i is a prescribed body force.

In addition we have the balance of mass equation (3.6). With π_{Ai} being the Piola-Kirchoff stress tensor and $F_{iA} = x_{i,A}$, the balance of angular momentum states

$$\pi \mathbf{F}^T = \mathbf{F} \pi^T.$$

One also requires an equation for the balance of energy in the body and this is

$$\rho_0 \dot{\varepsilon} = \pi_{Ai} \dot{F}_{iA} - q_{A,A} + \rho_0 r. \tag{3.9}$$

In this equation ε, q_A and r are, respectively, the internal energy function, the heat flux vector, and the externally supplied heat supply function. An explanation of equation (3.9) in terms of the physics of the global body is provided in the context of a thermoelastic body with a single distribution of voids in section 3.3.

To complete our thermodynamic development we require a suitable entropy inequality. We here use the Clausius-Duhem inequality

$$\rho_0 \dot{\eta} \geq -\left(\frac{q_A}{\theta}\right)_{,A} + \frac{\rho_0 r}{\theta}, \tag{3.10}$$

where η is the specific entropy function.

3.2.1 Thermodynamic Restrictions

We define a thermoelastic body to be one which has as constitutive variables the set

$$\Sigma = \{F_{iA}, \theta, \theta_{,A}, X_K\}. \tag{3.11}$$

The constitutive theory for a thermoelastic body assumes that the internal energy ε, the Piola-Kirchoff stress tensor π_{Ai}, the heat flux q_A, and the specific entropy depend on the variables in the list Σ, i.e.

$$\varepsilon = \varepsilon(\Sigma), \qquad \pi_{Ai} = \pi_{Ai}(\Sigma), \qquad q_A = q_A(\Sigma) \qquad \text{and} \qquad \eta = \eta(\Sigma). \tag{3.12}$$

The next step is to introduce the Helmholtz free energy function ψ by the relation

$$\varepsilon = \psi + \eta\theta. \tag{3.13}$$

We now employ (3.13) in equation (3.9) to see that

$$\rho_0 \dot{\psi} + \rho_0 \dot{\eta} \theta + \rho_0 \eta \dot{\theta} - \pi_{Ai} \dot{F}_{iA} = -q_{A,A} + \rho_0 r. \tag{3.14}$$

Expand the q_A term in inequality (3.10) and substitute for the terms $-q_{A,A} + \rho_0 r$ which result by employing (3.14). In this manner from inequality (3.10) we may show that

$$-\rho_0(\dot{\psi} + \eta\dot{\theta}) - \frac{q_A \theta_{,A}}{\theta} + \pi_{Ai}\dot{F}_{iA} \geq 0. \tag{3.15}$$

Now, use (3.13) together with (3.12) and expand $\dot{\psi}$ to find

$$-\left(\rho_0 \frac{\partial \psi}{\partial F_{iA}} - \pi_{Ai}\right)\dot{F}_{iA} - \rho_0\left(\frac{\partial \psi}{\partial \theta} + \eta\right)\dot{\theta} - \rho_0 \frac{\partial \psi}{\partial \theta_{,A}}\dot{\theta}_{,A} - \frac{q_A \theta_{,A}}{\theta} \geq 0. \tag{3.16}$$

Recall equations (3.12) and with this information one may recognise that the terms $\dot{F}_{iA}, \dot{\theta}$ and $\dot{\theta}_{,A}$ appear linearly in inequality (3.16). We may thus follow the procedure of Coleman & Noll [54] and assign an arbitrary value to each of these quantities in turn, ensuring that the conservation equations (3.7) and (3.9) are balanced by a suitable choice of the externally supplied body force f_i or heat supply r. It follows that inequality (3.16) may be violated unless the coefficients of $\dot{F}_{iA}, \dot{\theta}$ and $\dot{\theta}_{,A}$ are each identically zero. This leads to the deduction that the Helmholtz free energy function cannot depend on the temperature gradient $\theta_{,A}$, i.e.

$$\psi \neq \psi(\theta_{,A}) \tag{3.17}$$

and additionally

$$\pi_{Ai} = \rho_0 \frac{\partial \psi}{\partial F_{iA}} \quad \text{and} \quad \eta = -\frac{\partial \psi}{\partial \theta}. \tag{3.18}$$

Combining (3.17) with (3.18) and (3.13) we may further conclude that

$$\pi_{Ai} \neq \pi_{Ai}(\theta_{,A}), \quad \eta \neq \eta(\theta_{,A}), \quad \text{and} \quad \varepsilon \neq \varepsilon(\theta_{,A}).$$

The residual entropy inequality which remains from (3.16) is

$$-\frac{q_A \theta_{,A}}{\theta} \geq 0. \tag{3.19}$$

Thus, the equations of nonlinear thermoelasticity are (3.7) and (3.9) together with (3.18) and the constraint on the heat flux imposed by inequality (3.19). Upon employing relation (3.13) for the internal energy ε together with the expressions (3.18) for π_{Ai} and η one may reduce the energy balance equation (3.9) to

$$-\rho_0 \theta \frac{\partial}{\partial t}\left(\frac{\partial \psi}{\partial \theta}\right)\Big|_{\mathbf{X}} = -q_{A,A} + \rho_0 r. \tag{3.20}$$

Of course, equation (3.20) may be alternatively written as

$$\rho_0 \theta \dot{\eta} = -q_{A,A} + \rho_0 r. \tag{3.21}$$

3.2.2 *Linear Thermoelasticity*

We now include an exposition of the derivation of the equations of linear thermoelasticity from the fully nonlinear equations (3.7) and (3.21). Consider the reference state to be one where the temperature θ_0 is constant and let the displacement $u_i = x_i - X_i$ be small so that we add an ε and write $x_i = X_i + \varepsilon u_i$, where $0 < \varepsilon << 1$. We shall linearize about the isothermal state and put $\theta = \theta_0 + \varepsilon \hat{\theta}$. In the subsequent development we drop the hat on θ although it is to be understood that the ensuing term θ refers to the linearized perturbation temperature. We point out that one can allow a large non-isothermal deformation and then linearize about this state. This is very useful and allows one to have a deformation induced by a nonlinear temperature field. For a reader interested in this aspect we refer to the papers by Iesan [108, 110] or to the book by Iesan & Scalia [117].

Since (without the ε) $u_i = x_i - X_i$, it follows that $\ddot{x}_i = \ddot{u}_i$. We may relate the symmetric Cauchy stress t_{ij} to the Piola-Kirchoff stress π_{Ai} by the equation, see e.g. Iesan [111], p. 9, or Truesdell & Noll [220], p. 124,

$$Jt_{ij} = F_{iA}\pi_{Aj}\,, \tag{3.22}$$

where $J = \det \mathbf{F}$. If we momentarily denote the heat flux vector referred back to the reference configuration as Q_A as opposed to q_A and let q_i denote the heat flux vector in the current configuration then we also have, see Iesan [111], p. 9,

$$Jq_i = F_{iA}Q_A\,. \tag{3.23}$$

For equation (3.7) we transform the term $\pi_{Ai,A}$ into one involving the Cauchy stress in a linear theory, so we write

$$\frac{\partial \pi_{Ai}}{\partial X_A} = \frac{\partial \pi_{Ai}}{\partial x_j}F_{jA} = \frac{\partial \pi_{Ai}}{\partial x_j}(\delta_{jA} + \varepsilon u_{j,A}). \tag{3.24}$$

Further, we observe that

$$\begin{aligned}
\det \mathbf{F} &= \begin{vmatrix} 1+\varepsilon u_{1,1} & \varepsilon u_{1,2} & \varepsilon u_{1,3} \\ \varepsilon u_{2,1} & 1+\varepsilon u_{2,2} & \varepsilon u_{2,3} \\ \varepsilon u_{3,1} & \varepsilon u_{3,2} & 1+\varepsilon u_{3,3} \end{vmatrix} \\
&= 1 + \varepsilon \operatorname{div}\mathbf{u} + \varepsilon^2(u_{1,1}u_{2,2} + u_{1,1}u_{3,3} + u_{2,2}u_{3,3} - u_{1,3}u_{3,1} - u_{1,2}u_{2,1} - u_{2,3}u_{3,2}) \\
&\quad + \varepsilon^3(u_{1,1}u_{2,2}u_{3,3} + u_{1,2}u_{2,3}u_{3,1} + u_{1,3}u_{2,1}u_{3,2} \\
&\quad - u_{1,3}u_{3,1}u_{2,2} - u_{1,2}u_{2,1}u_{3,3} - u_{2,3}u_{3,2}u_{1,1})\,.
\end{aligned} \tag{3.25}$$

Then, employing (3.25), we may derive from (3.22)

$$\big(1 + O(\varepsilon)\big)t_{ij} = \pi_{Aj}(\delta_{iA} + \varepsilon u_{i,A})$$

and so to $O(1)$ one finds

$$t_{ij} \equiv \pi_{ij} \tag{3.26}$$

in the linearization. From (3.24) in linearizing we discard the ε term and see that

$$\frac{\partial \pi_{Ai}}{\partial X_A} \equiv \frac{\partial \pi_{Ai}}{\partial x_j}\delta_{jA} \equiv \frac{\partial t_{ji}}{\partial x_j}\,. \tag{3.27}$$

Hence, employing (3.26) and (3.27) in (3.7) we obtain the linearized momentum equation

$$\rho_0 \ddot{u}_i = t_{ji,j} + \rho_0 f_i\,. \tag{3.28}$$

In a similar manner, from (3.23) one finds

$$\det(\mathbf{I} + \varepsilon \nabla \mathbf{u})q_i = (\delta_{iA} + \varepsilon u_{i,A})Q_A$$

and so $q_i \equiv Q_i$ in the linear approximation. We may then write the linearized version of equation (3.21) as

$$\rho_0 \theta_0 \dot{\eta} = -q_{i,i} + \rho_0 r. \tag{3.29}$$

For a linearized theory we write the Helmholtz free energy function as

$$\rho_0 \psi = \frac{1}{2} a_{ijkh} e_{ij} e_{kh} - \frac{1}{2} a_0 \theta^2 - a_{ij} \theta e_{ij}, \tag{3.30}$$

where a_{ijkh}, a_0 and a_{ij} may depend on $\mathbf{x}$, and

$$e_{ij} = \frac{1}{2}(u_{i,j} + u_{j,i}).$$

We require the symmetries $a_{ijkh} = a_{khij} = a_{jikh}$. The analogous expressions to (3.18) are $t_{ji} = \rho_0 \partial \psi / \partial e_{ij}$ and $\eta = -\partial \psi / \partial \theta$ so one finds

$$\begin{aligned} t_{ij} &= a_{ijkh} e_{kh} - a_{ij} \theta \\ &= a_{ijkh} u_{k,h} - a_{ij} \theta \end{aligned} \tag{3.31}$$

and

$$\rho_0 \eta = a_0 \theta + a_{ij} e_{ij}. \tag{3.32}$$

We write q_i in the form of Fourier's law as

$$q_i = -k_{ij} \theta_{,j}, \tag{3.33}$$

where $k_{ij} = k_{ji}$ and from (3.19), $k_{ij} \xi_i \xi_j \geq 0$ for all ξ.

Thus, from (3.28), (3.21), (3.31), (3.32) and (3.33) the equations of linear thermoelasticity may be written as

$$\begin{aligned} \rho_0 \ddot{u}_i &= (a_{ijkh} u_{k,h})_{,j} - (a_{ij} \theta)_{,j} + \rho_0 f_i, \\ \rho_0 \theta_0 a_0 \dot{\theta} &= -\rho_0 \theta_0 a_{ij} \dot{u}_{i,j} + (k_{ij} \theta_{,j})_{,i} + \rho_0 r. \end{aligned} \tag{3.34}$$

It is worth pointing out that to interpret the nonlinear equation (3.7) correctly one needs to expand the term $\pi_{Ai,A}$ as

$$\pi_{Ai,A} = \frac{\partial \pi_{Ai}}{\partial F_{jB}} F_{jB,A} + \frac{\partial \pi_{Ai}}{\partial \theta} \theta_{,A} + \left. \frac{\partial \pi_{Ai}}{\partial X_A} \right|_{\mathbf{F},\theta} \tag{3.35}$$

where we allow for the fact that the coefficients of terms in π_{Ai} may depend on X_A. This may be understood with the aid of the linear theory since we have

$$\begin{aligned} \frac{\partial t_{ji}}{\partial x_j} &= (a_{ijkh} u_{k,h})_{,j} - (a_{ij} \theta)_{,j} \\ &= a_{ijkh} u_{k,hj} - a_{ij} \theta_{,j} + a_{ijkh,j} u_{k,h} - a_{ij,j} \theta. \end{aligned} \tag{3.36}$$

The first two terms on the right of (3.36) correspond to the first two terms on the right of (3.35) whereas the last two terms on the right of (3.36) correspond to the last term on the right of (3.35), the term arising from any inhomogeneity in ψ.

3.3 Single Porosity with Voids

The balance equations for a continuous body containing voids are given by Goodman & Cowin [94]. We use the equations as given by Nunziato & Cowin [165] since these are appropriate for an elastic body.

The key thing is to assume that there is a distribution of voids throughout the body B. If $\gamma(\mathbf{X},t)$ denotes the density of the elastic matrix, then the mass density $\rho(\mathbf{X},t)$ of B has form

$$\rho = \nu\gamma \tag{3.37}$$

where $0 < \nu \leq 1$ is a volume distribution function with $\nu = \nu(\mathbf{X},t)$. Since the density or void distribution in the reference configuration can be different we also have

$$\rho_0 = \nu_0\gamma_0$$

where ρ_0, γ_0, ν_0 are the equivalent functions to ρ, γ, ν, but in the reference configuration.

The first balance law is the balance of mass

$$\rho|\det\mathbf{F}| = \rho_0 .$$

With π_{Ai} being the Piola-Kirchoff stress tensor and $F_{iA} = x_{i,A}$ as before, the balance of angular momentum states

$$\pi\mathbf{F}^T = \mathbf{F}\pi^T .$$

The balance of linear momentum has form

$$\rho_0\ddot{x}_i = \pi_{Ai,A} + \rho_0 f_i, \tag{3.38}$$

f_i being an external body force. The balance law for the voids distribution is

$$\rho_0 k\ddot{\nu} = h_{A,A} + g + \rho_0\ell, \tag{3.39}$$

where k is an inertia coefficient, h_A is a stress vector, g is an intrinsic body force (giving rise to void creation/extinction inside the body), and ℓ is an external void body force. Actually, Nunziato & Cowin [165] allow the inertia coefficient k to depend on $\mathbf{X}$ and/or t, but, for simplicity, we follow Goodman & Cowin [94] and assume it to be constant.

The energy balance in the body may be expressed as

$$\rho_0\dot{\varepsilon} = \pi_{Ai}\dot{F}_{iA} + h_A\dot{\nu}_{,A} - g\dot{\nu} - q_{A,A} + \rho_0 r, \tag{3.40}$$

where ε, q_A and r are, respectively, the internal energy function, the heat flux vector, and the externally supplied heat supply function. To understand equation (3.40) we may integrate it over a fixed body B, integrate by parts, and use the divergence theorem to see that

$$\frac{d}{dt}\int_B \rho_0 \varepsilon dV + \int_B (g\dot{\nu} + h_{A,A}\dot{\nu})dV = \int_B \pi_{Ai}\dot{F}_{iA}dV - \oint_{\partial B} q_A N_A dS + \int_B \rho_0 r dV,$$

where ∂B is the boundary of B. Employing (3.39) with $\ell = 0$ we may rewrite the above as

$$\frac{d}{dt}\int_B (\rho_0 \varepsilon + \frac{\rho_0 k}{2}\dot{\nu}^2)dV = \int_B \pi_{Ai}\dot{F}_{iA}dV - \oint_{\partial B} q_A N_A dS + \int_B \rho_0 r dV.$$

In this form we recognise the equation as an energy balance equation with a term added due to the kinetic energy of the voids. In fact, Iesan [111], pp. 3–5, shows how one may begin with a conservation of energy law for an arbitrary sub-body of a continuous medium with voids, and then derive equations (3.38), (3.39) and (3.40) from the initial energy balance equation.

It is usual in continuum thermodynamics to also introduce an entropy inequality. We use the Clausius-Duhem inequality

$$\rho_0 \dot{\eta} \geq -\left(\frac{q_A}{\theta}\right)_{,A} + \frac{\rho_0 r}{\theta}, \tag{3.41}$$

where η is the specific entropy function. Observe that the sign of the first term on the right of (3.41) is different from that of Nunziato & Cowin [165]. (One could use a more sophisticated entropy inequality where q_A/θ is replaced by a general entropy flux $\mathbf{k}$, as in Goodman & Cowin [94], but the above is sufficient for our purpose.)

3.3.1 Thermodynamic restrictions

We consider an elastic body containing voids to be one which has as constitutive variables the set

$$\Sigma = \{\nu_0, \nu, F_{iA}, \theta, \theta_{,A}, \nu_{,A}, X_A\}. \tag{3.42}$$

Thus, the constitutive theory assumes

$$\begin{aligned} \varepsilon &= \varepsilon(\Sigma), & \pi_{Ai} &= \pi_{Ai}(\Sigma), & q_A &= q_A(\Sigma), \\ \eta &= \eta(\Sigma), & h_A &= h_A(\Sigma), & g &= g(\Sigma). \end{aligned} \tag{3.43}$$

This is different from Nunziato & Cowin [165] who regard η as the independent variable rather than θ and they also assume $q_A = 0$. The development of this theory is also described by Ciarletta & Iesan [50] and by Iesan [111]. For the case where $\dot{\nu}$ is also included in the constitutive list we refer to Straughan [202], pp. 310, 311.

To proceed we introduce the Helmholtz free energy function ψ in the manner

$$\varepsilon = \psi + \eta\theta. \tag{3.44}$$

Next, (3.40) is employed to remove the terms $-q_{A,A} + \rho_0 r$ from inequality (3.41) and then utilize (3.44) to rewrite (3.41) as

$$-\rho_0(\dot{\psi} + \eta\dot{\theta}) - \frac{q_A\theta_{,A}}{\theta} + \pi_{Ai}\dot{F}_{iA} + h_A\dot{\nu}_{,A} - g\dot{\nu} \geq 0. \tag{3.45}$$

The chain rule is used together with (3.43) to expand $\dot{\psi}$ and then (3.45) may be written as

$$\begin{aligned} &-\left(\rho_0\frac{\partial\psi}{\partial\nu} + g\right)\dot{\nu} - \frac{q_A\theta_{,A}}{\theta} - \left(\rho_0\frac{\partial\psi}{\partial F_{iA}} - \pi_{Ai}\right)\dot{F}_{iA} \\ &-\left(\rho_0\frac{\partial\psi}{\partial\theta} + \rho_0\eta\right)\dot{\theta} - \left(\rho_0\frac{\partial\psi}{\partial\nu_{,A}} - h_A\right)\dot{\nu}_{,A} - \rho_0\frac{\partial\psi}{\partial\theta_{,A}}\dot{\theta}_{,A} \geq 0. \end{aligned} \tag{3.46}$$

The next step is to observe that $\dot{F}_{iA}, \dot{\theta}, \dot{\theta}_{,A}, \dot{\nu}_{,A}$ and $\dot{\nu}$ appear linearly in inequality (3.46). We may then follow the procedure of Coleman & Noll [54] and assign an arbitrary value to each of these quantities in turn, balancing equations (3.38), (3.39) and (3.40) by a suitable choice of the externally supplied functions f_i, ℓ and r. We may in this manner violate inequality (3.46) unless the coefficients of $\dot{F}_{iA}, \dot{\theta}, \dot{\theta}_{,A}, \dot{\nu}_{,A}$ and $\dot{\nu}$ are each identically zero. Hence, we deduce that

$$\psi \neq \psi(\theta_{,A}),$$

$$h_A = \rho_0\frac{\partial\psi}{\partial\nu_{,A}} \quad \Rightarrow \quad h_A \neq h_A(\theta_{,A}), \tag{3.47}$$

$$\pi_{Ai} = \rho_0\frac{\partial\psi}{\partial F_{iA}} \quad \Rightarrow \quad \pi_{Ai} \neq \pi_{Ai}(,\theta_{,A}), \tag{3.48}$$

$$\eta = -\frac{\partial\psi}{\partial\theta} \quad \Rightarrow \quad \eta \neq \eta(\theta_{,A}), \tag{3.49}$$

$$g = -\rho_0\frac{\partial\psi}{\partial\nu} \tag{3.50}$$

and further

$$\varepsilon \neq \varepsilon(\theta_{,A}).$$

The residual entropy inequality, left over from (3.46), which must hold for all motions is

$$-\frac{q_A\theta_{,A}}{\theta} \geq 0.$$

Thus, to specify a material for an elastic body containing voids we have to postulate a suitable functional form for $\psi = \psi(\nu_0, \nu, F_{iA}, \theta, \nu_{,A}, X_A)$. Such a form is usually constructed with the aid of experiments.

3.3.2 Linearized Theory

Let us linearize about a state of constant temperature θ_0. The case of linearizing about a prestressed non-isothermal state is covered in Iesan [111], chapter 4.

We commence with the fully nonlinear equations

$$\begin{aligned}
\rho_0\ddot{x}_i &= \pi_{Ai,A} + \rho_0 f_i\,,\\
\rho_0 k\ddot{\nu} &= h_{A,A} + g + \rho_0\ell,\\
\rho_0\theta\dot{\eta} &= -q_{A,A} + \rho_0 r.
\end{aligned}\tag{3.51}$$

Equation $(3.51)_3$ is derived from the energy balance equation (3.40) by writing $\varepsilon = \psi + \eta\theta$, expanding $\dot{\psi}$, and employing the relations (3.47) - (3.50).

In a similar manner to the linearization process performed in section 3.2.2 we may derive linearized equations governing the variables u_i, ν and θ of form

$$\begin{aligned}
\rho_0\ddot{u}_i &= t_{ji,j} + \rho_0 f_i\,,\\
\rho_0 k\ddot{\nu} &= h_{i,i} + g + \rho_0\ell,\\
\rho_0\theta_0\dot{\eta} &= -q_{i,i} + \rho_0 r.
\end{aligned}\tag{3.52}$$

(See exercise 3.1.) In the linearized system the following equations hold

$$t_{ji} = \rho_0\frac{\partial\psi}{\partial e_{ij}}\,,\qquad h_i = \rho_0\frac{\partial\psi}{\partial\nu_{,i}}\,,\qquad \eta = -\frac{\partial\psi}{\partial\theta}\,,\qquad g = -\rho_0\frac{\partial\psi}{\partial\nu}\,.$$

For the linear theory with a centrosymmetric body we choose

$$\begin{aligned}
\rho_0\psi =& \frac{1}{2}a_{ijkh}e_{ij}e_{kh} - a_{ij}e_{ij}\theta - \frac{1}{2}a\theta^2\\
&+ \frac{1}{2}\alpha_{ij}\nu_{,i}\nu_{,j} + b_{ij}\nu e_{ij} + \frac{1}{2}\xi\nu^2 - m\theta\nu.
\end{aligned}$$

This gives rise to the constitutive equations

$$\begin{aligned}
t_{ij} &= a_{ijkh}u_{k,h} - a_{ij}\theta + b_{ij}\nu,\\
h_i &= \alpha_{ij}\nu_{,j}\,,\\
\rho_0\eta &= a_{ij}u_{i,j} + a\theta + m\nu,\\
g &= -\xi\nu + m\theta - b_{ij}e_{ij}\,,\\
q_i &= -\theta_0 k_{ij}\theta_{,j}\,,
\end{aligned}\tag{3.53}$$

where we have added θ_0 to the thermal conductivity k_{ij} simply for convenience.

Employing equations (3.53) in the linear balance laws (3.52) one obtains the evolutionary equations governing the motion of an anisotropic linear elastic solid containing a single distribution of voids as

$$\begin{aligned}
\rho_0\ddot{u}_i &= (a_{ijkh}u_{k,h})_{,j} - (a_{ij}\theta)_{,j} + (b_{ij}\nu)_{,j} + \rho_0 f_i\,,\\
\rho_0 k\ddot{\nu} &= (\alpha_{ij}\nu_{,j})_{,i} - \xi\nu + m\theta - b_{ij}u_{i,j} + \rho_0\ell,\\
a\dot{\theta} + m\dot{\nu} + a_{ij}\dot{u}_{i,j} &= (k_{ij}\theta_{,j})_{,i} + \frac{\rho_0}{\theta_0}r\,,
\end{aligned} \tag{3.54}$$

where f_i, ℓ and r are prescribed functions.

3.4 Double Porosity with Voids

In this section we describe the theory of Iesan & Quintanilla [115] who generalized the theory described in section 3.3 to the case where there are two voids distributions, one denoted by $\nu(\mathbf{X},t)$ which denotes the macro distribution of voids, and a second denoted by $\omega(\mathbf{X},t)$ which accounts for the micro distribution of voids. We use a different notation to that of Iesan & Quintanilla [115], one which is in keeping with the notation used elsewhere in this book.

The basic nonlinear equations of Iesan & Quintanilla [115] are the momentum equation

$$\rho_0\ddot{x}_i = \pi_{Ai,A} + \rho_0 f_i\,, \tag{3.55}$$

where the notation is as in sections 3.2, 3.3. In addition, there are now governing equations for both ν and ω and these have form

$$\kappa_1\ddot{\nu} = h_{A,A} + g + \rho_0\ell_1\,, \tag{3.56}$$

and

$$\kappa_2\ddot{\omega} = j_{A,A} + h + \rho_0\ell_2\,, \tag{3.57}$$

where ℓ_1 and ℓ_2 are externally supplied functions, while the fluxes h_A and j_A together with the intrinsic equilibrated body forces, g and h, are functions of the constitutive variables. Furthermore, one requires an energy balance equation and for Iesan & Quintanilla [115] this has form

$$\rho_0\dot{\varepsilon} = \pi_{Ai}\dot{F}_{iA} + h_A\dot{\nu}_{.A} - g\dot{\nu} + j_A\dot{\omega}_{,A} - h\dot{\omega} - q_{A,A} + \rho_o r, \tag{3.58}$$

where the notation is as in section 3.3, apart from new notation introduced in this section.

Iesan & Quintanilla [115] also employ the Clausius-Duhem inequality, which for ease in exposition we rewrite here as

$$\rho_0\dot{\eta} \geq -\left(\frac{q_A}{\theta}\right)_{,A} + \frac{\rho_0 r}{\theta}\,. \tag{3.59}$$

The first term on the right of inequality (3.59) is expanded and then the terms $-q_{A,A} + \rho_0 r$ are substituted using equation (3.58). Again the Helmholtz free energy function is introduced as $\psi = \varepsilon - \eta\theta$ and the entropy inequality may be rearranged to the form

$$-\rho_0(\dot{\psi}+\eta\dot{\theta})-\frac{q_A\theta_{,A}}{\theta}+\pi_{Ai}\dot{F}_{iA}+h_A\dot{\nu}_{,A}-g\dot{\nu}+j_A\dot{\omega}_{,A}-h\dot{\omega}\geq 0. \tag{3.60}$$

The set of constitutive variables chosen by Iesan & Quintanilla [115] is given by the list

$$\Sigma=\{F_{iA},\theta,\theta_{,A},\nu,\omega,\nu_{,A},\omega_{,A},X_A\}. \tag{3.61}$$

For constitutive theory Iesan & Quintanilla [115] assume that

$$\varepsilon,\eta,\pi_{Ai},q_A,h_A,j_A,g \quad \text{and} \quad h \tag{3.62}$$

are functions of the variables in the list Σ. Recalling $\psi=\varepsilon-\eta\theta$, $\dot{\psi}$ is now expanded in terms of the variables in (3.61) in inequality (3.60) to find

$$\begin{aligned}
&-\left(\rho_0\frac{\partial\psi}{\partial F_{iA}}-\pi_{Ai}\right)\dot{F}_{iA}-\left(\rho_0\frac{\partial\psi}{\partial\nu}+g\right)\dot{\nu}-\left(\rho_0\frac{\partial\psi}{\partial\omega}+h\right)\dot{\omega}\\
&-\left(\rho_0\frac{\partial\psi}{\partial\theta}+\rho_0\eta\right)\dot{\theta}-\rho_0\frac{\partial\psi}{\partial\theta_{,A}}\dot{\theta}_{,A}-\left(\rho_0\frac{\partial\psi}{\partial\nu_{,A}}-h_A\right)\dot{\nu}_{,A}\\
&-\left(\rho_0\frac{\partial\psi}{\partial\omega_{,A}}-j_A\right)\dot{\omega}_{,A}-\frac{q_A\theta_{,A}}{\theta}\geq 0.
\end{aligned} \tag{3.63}$$

Now recall the functional dependence in (3.62) and the composition of the list Σ. Due to the functions in Σ it follows that the terms $\dot{F}_{iA},\dot{\theta},\dot{\theta}_{,A},\dot{\nu},\dot{\omega},\dot{\nu}_{,A}$ and $\dot{\omega}_{,A}$ appear linearly in inequality (3.63). We may then follow the procedure of [54] and assign an arbitrary value to each of these quantities in turn, balancing equations (3.55) - (3.58) by a suitable choice of the externally supplied functions f_i,ℓ_1,ℓ_2 and r. We may in this manner violate inequality (3.63) unless the coefficients of $\dot{F}_{iA},\dot{\theta},\dot{\theta}_{,A},\dot{\nu},\dot{\omega},\dot{\nu}_{,A}$ and $\dot{\omega}_{,A}$ are each identically zero. Thus, one deduces ψ does not depend on $\theta_{,A}$. Moreover, one deduces the relations

$$\begin{aligned}
\pi_{Ai}&=\rho_0\frac{\partial\psi}{\partial F_{iA}}, & \eta&=-\frac{\partial\psi}{\partial\theta}, & g&=-\rho_0\frac{\partial\psi}{\partial\nu},\\
h_A&=\rho_0\frac{\partial\psi}{\partial\nu_{,A}}, & h&=-\rho_0\frac{\partial\psi}{\partial\omega}, & j_A&=\rho_0\frac{\partial\psi}{\partial\omega_{,A}}.
\end{aligned} \tag{3.64}$$

The residual entropy inequality from (3.63) is

$$-\frac{q_A\theta_{,A}}{\theta}\geq 0. \tag{3.65}$$

The energy equation (3.58) may be simplified using (3.64) to see that it has form

$$\rho_0\theta\dot{\eta}=-q_{A,A}+\rho_0 r. \tag{3.66}$$

The nonlinear system of governing equations for an elastic body with a double void distribution therefore consists of equations (3.55), (3.56), (3.57) and (3.66). The heat flux q_A is, in general, a function of the constitutive list (3.61).

3.4.1 Linearized Equations

One may proceed as in section 3.3 to carry out a linearization procedure linearizing about an isothermal state with constant temperature θ_0, see exercise 3.2. The details follow Iesan & Quintanilla [115]. To derive an anisotropic linear theory one selects, see Iesan & Quintanilla [115] equation (32)

$$\begin{aligned}\rho_0\psi =&\frac{1}{2}a_{ijkh}e_{ij}e_{kh} - a_{ij}\theta e_{ij} + b_{ij}\nu e_{ij} + d_{ij}\omega e_{ij}\\&+\frac{1}{2}\alpha_{ij}\nu_{,i}\nu_{,j} + \beta_{ij}\nu_{,i}\omega_{,j} + \frac{1}{2}\gamma_{ij}\omega_{,i}\omega_{,j} + \frac{\alpha_1}{2}\nu^2\\&+\frac{\alpha_2}{2}\omega^2 + \alpha_3\nu\omega - \gamma_1\nu\theta - \gamma_2\omega\theta - \frac{a}{2}\theta^2.\end{aligned} \tag{3.67}$$

Then the constitutive functions are given as

$$\begin{aligned}t_{ij} =&a_{ijkh}u_{k,h} - a_{ij}\theta + b_{ij}\nu + d_{ij}\omega,\\ \eta =&a\theta + \gamma_1\nu + \gamma_2\omega + a_{ij}e_{ij},\\ h_i =&\alpha_{ij}\nu_{,j} + \beta_{ij}\omega_{,j},\\ j_i =&\gamma_{ij}\omega_{,j} + \beta_{ij}\nu_{,j},\\ g =&-\alpha_1\nu - \alpha_3\omega + \gamma_1\theta - b_{ij}u_{i,j},\\ h =&-\alpha_2\omega - \alpha_3\nu + \gamma_2\theta - d_{ij}u_{i,j}.\end{aligned} \tag{3.68}$$

If the heat flux q_i satisfies a Fourier law then one may take $q_i = -\theta_0 k_{ij}\theta_{,j}$ where k_{ij} is a non-negative definite tensor. The linearized equations thus take the form

$$\begin{aligned}&\rho_0\ddot{u}_i = (a_{ijkh}u_{k,h})_{,j} - (a_{ij}\theta)_{,j} + (b_{ij}\nu)_{,j} + (d_{ij}\omega)_{,j} + \rho_0 f_i,\\&\kappa_1\ddot{\nu} = (\alpha_{ij}\nu_{,j})_{,i} + (\beta_{ij}\omega_{,j})_{,i} - \alpha_1\nu - \alpha_3\omega + \gamma_1\theta - b_{ij}u_{i,j} + \rho_0\ell_1,\\&\kappa_2\ddot{\omega} = (\gamma_{ij}\omega_{,j})_{,i} + (\beta_{ij}\nu_{,j})_{,i} - \alpha_2\omega - \alpha_3\nu + \gamma_2\theta - d_{ij}u_{i,j} + \rho_0\ell_2,\\&a\dot{\theta} + \gamma_1\dot{\nu} + \gamma_2\dot{\omega} + a_{ij}\dot{u}_{i,j} = (k_{ij}\theta_{,j})_{,i} + \frac{\rho_0}{\theta_0}r.\end{aligned} \tag{3.69}$$

3.5 Exercises

Exercise 3.1. Justify the linearization process leading to equations (3.52) from the nonlinear equations (3.51). Use a process involving $x_i = X_i + \varepsilon u_i$, $\theta = \theta_0 + \varepsilon\hat{\theta}$,

$v = v_0 + \varepsilon\hat{v}$, $Jt_{ij} = F_{iA}\pi_{Aj}$, $Jq_i = F_{iA}Q_A$, $Jh_i = F_{iA}H_A$, and drop the hats on θ and v.

Exercise 3.2. Justify the linearization process in section 3.4.1. Derive the equations analogous to (3.52) which arise from the nonlinear equations (3.55), (3.56), (3.57) and (3.66). You should use a process involving $x_i = X_i + \varepsilon u_i$, $\theta = \theta_0 + \varepsilon\hat{\theta}$, $v = v_0 + \varepsilon\hat{v}$, $\omega = \omega_0 + \varepsilon\hat{\omega}$, $Jt_{ij} = F_{iA}\pi_{Aj}$, $Jq_i = F_{iA}Q_A$, $Jh_i = F_{iA}H_A$, $Jj_i = F_{iA}J_A$, and drop the hats on θ, v and ω.

Exercise 3.3. Consider the boundary-initial value problem, $\mathscr{P}$, for the linear thermoelasticity equations (3.34), where u_i and θ are prescribed on the boundary Γ, and $u_i, \dot{u}_i$ and θ are given as initial data. Let a_{ijkh} be such that

$$a_{ijkh}\xi_{ij}\xi_{kh} \geq 0 \qquad \forall\, \xi_{ij}, \tag{3.70}$$

and suppose k_{ij}, a_{ij} are symmetric with

$$k_{ij}\xi_i\xi_j \geq 0 \qquad \forall\, \xi_i\,.$$

Show by an energy method that the solution to $\mathscr{P}$ is unique.
Hint. Write the system as

$$\begin{aligned} \rho_0\ddot{u}_i &= (a_{ijkh}u_{k,h})_{,j} - (a_{ij}\theta)_{,j} + \rho_0 f_i\,, \\ a_0\dot{\theta} &= -a_{ij}\dot{u}_{i,j} + (K_{ij}\theta_{,j})_{,i} + \hat{r}\,, \end{aligned} \tag{3.71}$$

where $K_{ij} = k_{ij}/\rho_0\theta_0$ and $\hat{r} = r/r_0$.

Exercise 3.4. Consider the equivalent boundary-initial value problem, $\mathscr{PE}$, to that of exercise 3.3 but where now Ω is a domain exterior to a bounded domain $\Omega_0 \subset \mathbb{R}^3$. Given u_i, θ on the inner boundary Γ and $u_i, \dot{u}_i, \theta$ on $\Omega \times \{t = 0\}$ show that the solution to $\mathscr{PE}$ is unique supposing as $r \to \infty$,

$$|\dot{u}_i|, |u_{i,j}|, |\theta|, |\theta_{,i}| \leq ke^{\lambda r}\,.$$

Hint. Multiply the difference equation for $(3.71)_1$ by $g\dot{u}_i$ and integrate over Ω. Multiply the difference equation for $(3.71)_2$ by $g\theta$ and integrate over Ω. You should obatin the equation

$$\begin{aligned} &\frac{d}{dt}\frac{1}{2}\left(\int_\Omega g\rho_0\dot{u}_i\dot{u}_i\,dx + \int_\Omega g a_{ijkh}u_{i,j}u_{k,h}\,dx + \int_\Omega g a_0\theta^2\,dx\right) \\ &\quad + \int_\Omega gK_{ij}\theta_{,i}\theta_{,j}\,dx = \delta\int_\Omega g\frac{x_j}{r}\,a_{ijkh}\dot{u}_i u_{k,h}\,dx \\ &\quad - \int_\Omega g\frac{x_i}{r}\,\dot{u}_i\theta\,dx + \delta\int_\Omega g\frac{x_i}{r}\,K_{ij}\theta\theta_{,j}\,dx\,. \end{aligned}$$

Balance the $\theta_{,j}$ terms on the right with those on the left and then follow section 1.3.5.

Exercise 3.5. Consider the equivalent boundary-initial value problem to $\mathscr{P}$ in exercise 3.3 but do not require condition (3.70). Require instead only the symmetry conditions

$$a_{ijkh} = a_{khij} = a_{jikh}\,, \tag{3.72}$$

and suppose k_{ij}, a_{ij} are symmetric with

$$k_{ij}\xi_i\xi_j \geq 0 \qquad \forall \xi_i\,.$$

By using a logarithmic convexity method show that the solution to $\mathscr{P}$ is unique.
Hint. Consider the function

$$F(t) = < \rho_0 u_i u_i > + \int_0^t < K_{ij}\eta_{,i}\eta_{,j} > ds$$

where

$$\eta(\mathbf{x},t) = \int_0^t \theta(\mathbf{x},s)ds$$

and where u_i, θ denote difference variables, (cf. Straughan [201]).

Exercise 3.6. Consider system (3.71) with $f_i = 0, \hat{r} = 0$. If we discard the acceleration term $\rho_0 \ddot{u}_i$ the resulting equations are known as the quasi-equilibrium equations of linear thermoelasticity, namely,

$$\begin{aligned} 0 &= (a_{ijkh}u_{k,h})_{,j} - (a_{ij}\theta)_{,j}\,, \\ a\dot{\theta} &= -a_{ij}\dot{u}_{i,j} + (K_{ij}\theta_{,j})_{,i}\,, \end{aligned} \tag{3.73}$$

where $a(\mathbf{x})$ is employed rather than $a_0(\mathbf{x})$ as in (3.71).

Suppose a_{ij} and K_{ij} are symmetric and a_{ijkh} satisfy

$$a_{ijkh} = a_{khij} = a_{jikh}\,.$$

Suppose further

$$a_{ijkh}\xi_{ij}\xi_{kh} \geq a_0\xi_{ij}\xi_{ij}\,, \quad \forall\, \xi_{ij}\,, \qquad K_{ij}\xi_i\xi_j \geq k_0\xi_i\xi_i\,, \quad \forall \xi_i,$$

where a_0, k_0 are positive constants.

Let (3.73) be defined on $\Omega \times \{t > 0\}$ and suppose

$$|a_{ij}| \leq A, \qquad a_1 \geq a(\mathbf{x}) \geq \hat{a} > 0,$$

$\forall \mathbf{x} \in \Omega$, and for constants $\hat{a}, a_1$ and A.

Suppose that $u_i = 0, \theta = 0$ on Γ, the boundary of Ω. Deduce that the solution to the boundary-initial value problem for (3.73) with zero boundary conditions decays in a suitable sense.
Hint. Multiply $(3.73)_1$ by u_i, integrate over Ω, and integrate by parts to show

$$< a_{ijkh}u_{i,j}u_{k,h} > = < a_{ij}\theta u_{i,j} >\,.$$

Then use the inequalities given and the arithmetic-geometric mean inequality to show that

$$< a_{ijkh}u_{i,j}u_{k,h} > \leq \frac{A}{2\alpha\hat{a}} < a\theta^2 > + \frac{A\alpha}{2a_0} < a_{ijkh}u_{i,j}u_{k,h} >$$

for some constant $\alpha > 0$. Choose $\alpha = a_0/A$ to obtain

$$< a_{ijkh}u_{i,j}u_{k,h} > \leq \frac{A^2}{a_0\hat{a}} < a\theta^2 > . \tag{3.74}$$

Next, multiply $(3.73)_1$ by $\dot{u}_i$ and $(3.73)_2$ by θ, integrate over Ω with some integration by parts to deduce

$$\begin{aligned} \frac{d}{dt}\frac{1}{2} < a_{ijkh}u_{i,j}u_{k,h} > &= < a_{ij}\theta\dot{u}_{i,j} >, \\ \frac{d}{dt}\frac{1}{2} < a\theta^2 > &= - < K_{ij}\theta_{,i}\theta_{,j} > - < a_{ij}\theta\dot{u}_{i,j} > . \end{aligned}$$

Let

$$F(t) = \frac{1}{2} < a_{ijkh}u_{i,j}u_{k,h} > + \frac{1}{2} < a\theta^2 >$$

and then show that

$$\frac{dF}{dt} = - < K_{ij}\theta_{,i}\theta_{,j} > . \tag{3.75}$$

Now, show

$$< K_{ij}\theta_{,i}\theta_{,j} > \geq \frac{k_0\lambda_1}{a_1} < a\theta^2 >$$

where λ_1 is the Poincaré constant for Ω. Integrate (3.75) to find

$$F(t) = F(0) - \int_0^t < K_{ij}\theta_{,i}\theta_{,j} > ds.$$

Deduce that

$$G'(t) + \alpha G \leq F(0)$$

where $\alpha = 2k_0\lambda_1/a_0$ and

$$G(t) = \frac{1}{2}\int_0^t < a\theta^2 > ds.$$

Then show

$$G(t) \leq \frac{F(0)}{\alpha}(1 - exp(-\alpha t)), \qquad \forall t > 0. \tag{3.76}$$

Since (3.76) holds for all $t > 0$ deduce that $G \in L^1(0,\infty)$ i.e.

$$\frac{1}{2}\int_0^\infty < a\theta^2 > ds \leq \frac{F(0)}{\alpha} < \infty .$$

Further, use (3.74) to conclude

$$\int_0^t F\,ds \le \Big(1+\frac{A^2}{\hat{a}a_0}\Big)\frac{F(0)}{\alpha} < \infty.$$

Hence, $F \in L^1(0,\infty)$. Then use the facts that $F \ge 0$ and $F' \le 0$, to deduce $F \to 0$ as $t \to \infty$.

(Thus one obtains decay in time in the measures $< a\theta^2 >$ and $< a_{ijkh}u_{i,j}u_{k,h} >$.)

General decay results for the full thermoelastic system (3.34) are non-trivial and depend on the geometry of Ω, see Lebeau & Zuazua [150].

Exercise 3.7. In the isothermal case when viscoelastic effects are taken into account equations (3.54) are modified to include a dissipation term in $\dot{\nu}$. The relevant equations are then, see Ciarletta & Iesan [50], Cowin [57]

$$\begin{aligned} \rho\ddot{u}_i &= (a_{ijkh}u_{k,h})_{,j} + (b_{ij}\nu)_{,j} + \rho f_i, \\ \chi_1\ddot{\nu} &= (\alpha_{ij}\nu_{,j})_{,i} - \tau\dot{\nu} - b_{ij}u_{i,j} - \xi\nu + \rho\ell, \end{aligned} \tag{3.77}$$

where $\chi_1 = \rho k$ and $\tau > 0$ is a constant. Consider the boundary-initial value problem, $\mathscr{P}\mathscr{V}$, for (3.77) for which u_i, ν are given on Γ, and $u_i, \dot{u}_i, \nu$ and $\dot{\nu}$ are given in Ω at $t = 0$. Assume b_{ij}, α_{ij} are symmetric with α_{ij} non-negative and a_{ijkh} satisfy only the symmetry conditions (3.72). By using a logarithmic convexity method show that the solution to $\mathscr{P}\mathscr{V}$ is unique.
Hint. Employ a logarithmic convexity argument with the function

$$F(t) = < \rho u_i u_i > + < \chi_1 \nu^2 > + \int_0^t < \tau\nu^2 > ds$$

where u_i and ν represent the difference solution.

Exercise 3.8. Consider the boundary-initial value problem for the thermoviscoelastic equations (3.54), in which u_i, ν and θ are given on Γ and $u_i, \dot{u}_i, \nu, \dot{\nu}$ and θ are given on Ω at $t = 0$. Assume $a_{ij}, b_{ij}, \alpha_{ij}, k_{ij}$ are symmetric with α_{ij} and k_{ij} non-negative. Further assume a_{ijkh} satisfy only the symmetry conditions (3.72). Show that the solution to the above boundary-initial value problem is unique.
Hint. Use a logarithmic convexity argument with the function

$$F(t) = < \rho_0 u_i u_i > + < \chi_1 \nu^2 > + \int_0^t < k_{ij}\eta_{,i}\eta_{,j} > ds$$

where

$$\eta(\mathbf{x},t) = \int_0^t \theta(\mathbf{x},s)ds$$

and where u_i, ν and θ represent the difference solution.

Exercise 3.9. Consider the boundary-initial value problem, $\mathscr{P}$, for the thermoviscoelastic voids model which generalizes (3.54) and is governed by the equations

$$\begin{aligned}
&\rho \ddot{u}_i = (a_{ijkh}u_{k,h})_{,j} - (a_{ij}\theta)_{,j} + (b_{ij}\nu)_{,j} + \rho f_i, \\
&\chi_1 \ddot{\nu} = (\alpha_{ij}\nu_{,j})_{,i} - \tau\dot{\nu} - b_{ij}u_{i,j} + m\theta - \xi\nu + \rho\ell, \\
&a\dot{\theta} + m\dot{\nu} = (k_{ij}\theta_{,j})_{,i} - a_{ij}\dot{u}_{i,j} + \rho r.
\end{aligned} \tag{3.78}$$

Suppose u_i, ν and θ are given on Γ together with and $u_i, \dot{u}_i, \nu, \dot{\nu}$ and θ given on Ω when $t = 0$. Suppose the symmetries of exercise (3.7) hold. Show that a solution to $\mathscr{P}$ is unique.

Hint. Use a logarithmic convexity argument with the function

$$F(t) = <\rho u_i u_i> + <\chi_1 \nu^2> + \int_0^t <\tau \nu^2> ds + \int_0^t <k_{ij}\eta_{,i}\eta_{,j}> ds$$

where

$$\eta(\mathbf{x},t) = \int_0^t \theta(\mathbf{x},s) ds$$

and where u_i and ν represent the difference solution.

Exercise 3.10. For a thermoviscoelastic body containing voids Iesan [114], see also D'Apice & Chirita [60], produces the following system of governing equations

$$\begin{aligned}
&\rho \ddot{u}_i = (a_{ijkh}u_{k,h})_{,j} + (\hat{a}_{ijkh}\dot{u}_{k,h})_{,j} - (a_{ij}\theta)_{,j} + (b_{ij}\phi)_{,j} + (\hat{b}_{ij}\dot{\phi})_{,j}\,, \\
&\chi_1 \ddot{\phi} = (k_{ij}\phi_{,j})_{,i} + (\hat{k}_{ij}\dot{\phi}_{,j})_{,i} - b_{ij}u_{i,j} - \tilde{b}_{ij}\dot{u}_{i,j} - \xi\phi - \tau\dot{\phi} + (\tau_{ij}\theta_{,j})_{,i} + m\theta, \\
&a\dot{\theta} + m\dot{\phi} = (\alpha_{ij}\theta_{,j})_{,i} - a_{ij}\dot{u}_{i,j} + (\zeta_{ij}\dot{\phi}_{,j})_{,i}\,,
\end{aligned} \tag{3.79}$$

where ϕ is now the voids distribution and $b_{ij}, \hat{b}_{ij}, a_{ij}, k_{ij}, \hat{k}_{ij}, \tilde{b}_{ij}, \tau_{ij}, \alpha_{ij}$ and ζ_{ij} are all symmetric tensors. Let the boundary - initial value problem for (3.79) with u_i, ϕ defined on Γ and $u_i, \dot{u}_i, \phi, \dot{\phi}$ and θ defined on $\Omega \times \{t = 0\}$ be denoted by $\mathscr{P}$. Let all coefficients depend on $\mathbf{x}$ and suppose

$$|\tau_{ij} + \zeta_{ij}| \le M, \qquad |\hat{b}_{ij} + \tilde{b}_{ij}| \le A,$$

for constants $A, M > 0$. Suppose

$$\begin{aligned}
&\hat{k}_{ij}\xi_i\xi_j \ge k_0\xi_i\xi_i\,, \qquad \alpha_{ij}\xi_i\xi_j \ge \alpha_0\xi_i\xi_i\,, \\
&\hat{a}_{ijkh}\xi_{ij}\xi_{kh} \ge \hat{a}\xi_{ij}\xi_{ij}\,, \qquad \tau \ge \tau_0\,,
\end{aligned} \tag{3.80}$$

for all ξ_i, ξ_{ij} where $k_0, \alpha_0 \ge M/2 > 0$, $\hat{a} > 0$, $\tau_0 > 0$. In addition suppose the following inequality holds

$$\begin{aligned}
&\frac{1}{2} <a_{ijkh}\mu_{i,j}\mu_{k,h}> + <b_{ij}\mu_{i,j}\mu> + \frac{1}{2} <\xi\mu^2> \\
&\quad + \frac{1}{2} <k_{ij}\mu_{,i}\mu_{,j}> \ge a_0 <\mu_{i,j}\mu_{i,j}> + a_1\|\mu\|^2,
\end{aligned} \tag{3.81}$$

for all functions μ_i, μ which are zero on Γ, where $a_0 > 0, a_1 > 0$, $<\cdot>$ denotes integration over Ω and $\|\cdot\|$ denotes the norm on $L^2(\Omega)$.

Prove the solution to $\mathscr{P}$ is unique.

Hint. Employ an energy technique. Let u_i, ϕ and θ denote the difference of two solutions which satisfy (3.79) and satisfy the same boundary and initial conditions. Then multiply the difference equations for (3.79) by $\dot{u}_i$ for $(3.79)_1$, multiply by $\dot{v}$ for $(3.79)_2$ and multiply by θ for $(3.79)_3$. In this way derive the energy equation

$$\begin{aligned}\frac{dE}{dt} &+ < \hat{a}_{ijkh}\dot{u}_{i,j}\dot{u}_{k,h} > + < \hat{k}_{ij}\dot{\phi}_{,i}\dot{\phi}_{,j} > + \alpha_{ij}\theta_{,i}\theta_{,j} > \\ &+ < \tau\dot{\phi}^2 > = - < (\tilde{b}_{ij} + \hat{b}_{ij})\dot{u}_{i,j}\dot{\phi} > - < (\tau_{ij} + \zeta_{ij})\theta_{,i}\dot{\phi}_{,j} >\end{aligned} \tag{3.82}$$

where

$$\begin{aligned}E(t) =& \frac{1}{2} < \rho\dot{u}_i\dot{u}_i > + \frac{1}{2} < a_{ijkh}u_{i,j}u_{k,h} > + < b_{ij}\phi u_{i,j} > \\ &+ \frac{1}{2} < \chi_1\dot{\phi}^2 > + \frac{1}{2} < k_{ij}\phi_{,i}\phi_{,j} > + \frac{1}{2} < \xi\phi^2 > + \frac{1}{2} < a\theta^2 > .\end{aligned} \tag{3.83}$$

Integrate equation (3.82) from 0 to t and then involve (3.80) and (3.81) after using the arithmetic-geometric mean inequality on the right of (3.82). In this way derive the inequality

$$\begin{aligned}\frac{1}{2} < \rho\dot{u}_i\dot{u}_i > &+ a_0\|\nabla\mathbf{u}\|^2 + \frac{1}{2} < \chi_1\dot{\phi}^2 > \\ &+ \frac{1}{2} < a\theta^2 > + a_1\|\phi\|^2 \leq C\int_0^t \|\dot{\phi}\|^2 ds,\end{aligned} \tag{3.84}$$

where $C = (A^2/4\hat{a} - \tau_0)$. Suppose $\chi_1 \geq \chi^* > 0$, $a \geq a^* > 0$ and define $\mathscr{F}$ by

$$\mathscr{F} = \frac{1}{2} < \rho\dot{u}_i\dot{u}_i > + a_0\|\nabla\mathbf{u}\|^2 + \frac{\chi^*}{2}\|\dot{\phi}\|^2 > + \frac{1}{2} < a\theta^2 > + a_1\|\phi\|^2 .$$

Deduce that

$$\mathscr{F} \leq C\int_0^t \|\dot{\phi}\|^2 ds. \tag{3.85}$$

When $C \leq 0$ deduce uniqueness directly from (3.85). Further, when $C > 0$, deduce there is a constant $B > 0$ such that

$$\mathscr{F} \leq B\int_0^t \mathscr{F}\, ds.$$

Integrate this inequality with an integrating factor to deduce

$$G(t) \leq G(0)e^{Bt} = 0$$

where $G(t) = \int_0^t \mathscr{F}\, ds$. Thus deduce $\mathscr{F} \equiv 0$ and so $\phi \equiv 0, \theta \equiv 0$ and with the aid of Poincaré's inequality $u_i \equiv 0$.

Exercise 3.11. Consider the boundary-initial value problem of exercise 3.10 with $\hat{b}_{ij} = 0$, $\tilde{b}_{ij} = 0$, $\tau_{ij} = 0$ and $\zeta_{ij} = 0$. Suppose the remaining conditions of exercise

3.10 hold except that we do not require definiteness on a_{ijkh}, we require only that the symmetry conditions (3.72) hold. Show that a solution to the boundary-initial value problem then generated is unique.
Hint. Use a logarithmic convexity argument with the function F given by

$$F(t) = < \rho_0 u_i u_i > + < \chi_1 \phi^2 > + \int_0^t < \tau\phi^2 > ds + \int_0^t < \alpha_{ij}\eta_{,i}\eta_{,j} > ds$$
$$+ \int_0^t < \hat{a}_{ijkh} u_{i,j} u_{k,h} > ds + \int_0^t < \hat{k}_{ij}\phi_{,i}\phi_{,j} > ds$$

where

$$\eta(\mathbf{x},t) = \int_0^t \theta(\mathbf{x},s) ds$$

and where u_i, ϕ and θ represent the difference solution.

Exercise 3.12. If one were to combine the single voids system (3.77) with the Kelvin-Voigt elasticity equations, see (1.61), then one may derive a model for a viscoelastic body containing a single distribution of voids. The governing equations would be

$$\begin{aligned} \rho \ddot{u}_i &= (a_{ijkh} u_{k,h})_{,j} + (b_{ijkh} \dot{u}_{k,h})_{,j} + (b_{ij}\phi)_{,j} \\ \chi_1 \ddot{\phi} &= (a_{ij}\phi_{,j})_{,i} - \tau\dot{\phi} - b_{ij} u_{i,j} - \xi\phi , \end{aligned} \tag{3.86}$$

cf. exercise 3.11. A more complete model involving other dissipation effects is due to Iesan [114], see also exercise 3.10.

By using an argument similar to that of exercise 1.12, shows that a solution to the displacement boundary - initial value problem with zero boundary conditions will decay exponentially under suitable conditions on a_{ijkh} and b_{ijkh}.

Exercise 3.13. Consider the boundary-initial value problem $\mathscr{P}$ for equations (3.86) with

$$u_i(\mathbf{x},t) = u_i^B(\mathbf{x},t), \qquad \phi(\mathbf{x},t) = \phi^B(\mathbf{x},t), \qquad \text{on } \Gamma \times (0,T)$$

and

$$\begin{aligned} u_i(\mathbf{x},0) &= v_i(\mathbf{x}), & \dot{u}_i(\mathbf{x},0) &= w_i(\mathbf{x}), \\ \phi(\mathbf{x},0) &= \Phi(\mathbf{x}), & \dot{\phi}(\mathbf{x},0) &= \Psi(\mathbf{x}) \end{aligned}$$

for $u_i^B, \phi^B, v_i, w_i, \Phi$ and Ψ given functions. Show that a solution to $\mathscr{P}$ is unique when

$$a_{ijkh} = a_{jikh} = a_{khij} \tag{3.87}$$

with b_{ijkh} satsifying the same symmetries but being non-negative. The functions a_{ij}, b_{ij} are symmetric and a_{ij} is non-negative. The other coefficients are positive.
Hint. Use a logarithmic convexity argument with the function F given by

$$F(t) =< \rho_0 u_i u_i > + < \chi_1 \phi^2 > + \int_0^t < \tau\phi^2 > ds + \int_0^t < b_{ijkh} u_{i,j} u_{k,h} > ds.$$

Exercise 3.14. Consider the boundary-initial value problem $\mathscr{P}$ for equations (3.86) with

$$u_i(\mathbf{x},t) = u_i^B(\mathbf{x},t), \qquad \phi(\mathbf{x},t) = \phi^B(\mathbf{x},t), \qquad \text{on } \Gamma \times (0,T)$$

and

$$\begin{aligned} u_i(\mathbf{x},0) &= v_i(\mathbf{x}), & \dot{u}_i(\mathbf{x},0) &= w_i(\mathbf{x}), \\ \phi(\mathbf{x},0) &= \Phi(\mathbf{x}), & \dot{\phi}(\mathbf{x},0) &= \Psi(\mathbf{x}) \end{aligned}$$

for $u_i^B, \phi^B, v_i, w_i, \Phi$ and Ψ given functions. Show that a solution to $\mathscr{P}$ depends continuously upon the initial data when the coefficients in the equations are as in exercise 3.13.
Hint. Use a logarithmic convexity method as in section 1.4.5 or section 1.4.6, and let

$$\begin{aligned} F(t) = &< \rho_0 u_i u_i > + < \chi_1 \phi^2 > + \int_0^t < \tau\phi^2 > ds + \int_0^t < b_{ijkh} u_{i,j} u_{k,h} > ds \\ &+ (T-t)\big[< \tau\Phi^2 > + < b_{ijkh} v_{i,j} v_{k,h} >\big]. \end{aligned}$$

When $E(0) \leq 0$ show $\log F$ is a convex function of t. When $E(0) > 0$ put $J = F(t) + 2E(0)$ and show $H = \log J(t) + t^2$ is a convex function of t.

Chapter 4
Comparison of Porosity and Voids Theories

In chapter 2 we discussed several different models for the description of a double porosity elastic medium. Chapter 3 also derives another model for an elastic body containing two levels of porosity, a theory based on two voids distributions. In this chapter we shall compare the Svanadze model of chapter 3, see section 2.3, to that of Iesan & Quintanilla [115] as discussed in section 3.4.1. The comparison we make is based on the equations of a linearized theory and we use two methods to compare the solution behaviour. We firstly study the movement of an acceleration wave. The second method is to study how uniqueness follows from a logarithmic convexity argument when the elastic coefficients are not required to be sign definite.

We now present the basic equations to enable a comparison.

4.1 Governing Equations

Even though the book focusses on double porosity and triple porosity theories for an elastic material, to understand how the porosity and voids theories differ it is instructive to commence with equations for a linear elastic body with one set of pores, or one void distribution.

4.1.1 One Porosity Theory

When the anisotropic linearly elastic body possesses one set of pores then the governing equations with no cross inertia follow from (2.25) and (2.26) neglecting the γ and γ_{ij} terms. Thus, for one porosity the relevant equations are

B. Straughan, *Mathematical Aspects of Multi–Porosity Continua*, Advances in Mechanics and Mathematics 38, https://doi.org/10.1007/978-3-319-70172-1_4

$$\begin{aligned}\rho\ddot{u}_i &= \frac{\partial}{\partial x_j}\Big(a_{ijkh}\frac{\partial u_k}{\partial x_h}\Big) - \frac{\partial}{\partial x_j}(\beta_{ij}p) + \rho f_i\,,\\ \alpha\dot{p} &= \frac{\partial}{\partial x_i}\Big(k_{ij}\frac{\partial p}{\partial x_j}\Big) - \beta_{ij}\frac{\partial \dot{u}_i}{\partial x_j} + \rho s_1.\end{aligned} \tag{4.1}$$

In this section we shall suppose β_{ij} and k_{ij} are symmetric and the elastic coefficients satisfy the symmetries

$$a_{ijkh} = a_{khij} = a_{jikh}\,. \tag{4.2}$$

4.1.2 One Void Theory

When the elastic body has one void distribution the equations follow from (3.54) with θ set equal to zero and one neglects the θ equation. Thus, the appropriate equations are

$$\begin{aligned}\rho\ddot{u}_i &= \frac{\partial}{\partial x_j}\Big(a_{ijkh}\frac{\partial u_k}{\partial x_h}\Big) + \frac{\partial}{\partial x_j}(b_{ij}\nu) + \rho f_i\,,\\ \kappa_1\ddot{\nu} &= \frac{\partial}{\partial x_i}\Big(\alpha_{ij}\frac{\partial \nu}{\partial x_j}\Big) - b_{ij}\frac{\partial u_i}{\partial x_j} - \xi\nu + \rho\ell.\end{aligned} \tag{4.3}$$

Again, the elastic coefficients satisfy (4.2) and α_{ij} and b_{ij} are symmetric tensors.

4.1.3 Double Porosity Theory

The governing equations in this case are given by equations (2.25)-(2.27) where $\beta_{ij}, \gamma_{ij}, k_{ij}$ and m_{ij} are symmetric tensors and the elasticities satisfy (4.2). For completeness and ease of understanding the exposition in this section we record the equations here as

$$\begin{aligned}\rho\ddot{u}_i &= \frac{\partial}{\partial x_j}\Big(a_{ijkh}\frac{\partial u_k}{\partial x_h}\Big) - \frac{\partial}{\partial x_j}(\beta_{ij}p) - \frac{\partial}{\partial x_j}(\gamma_{ij}q) + \rho f_i\,,\\ \alpha\dot{p} &= \frac{\partial}{\partial x_i}\Big(k_{ij}\frac{\partial p}{\partial x_j}\Big) - \beta_{ij}\frac{\partial \dot{u}_i}{\partial x_j} - \gamma(p-q) + \rho s_1\,,\\ \beta\dot{q} &= \frac{\partial}{\partial x_i}\Big(m_{ij}\frac{\partial q}{\partial x_j}\Big) - \gamma_{ij}\frac{\partial \dot{u}_i}{\partial x_j} + \gamma(p-q) + \rho s_2\,.\end{aligned} \tag{4.4}$$

4.1.4 Double Voids Theory

When there are two void distributions present the governing equations arise from (3.69) by setting $\theta = 0$ and neglecting the θ equation. In this case the relevant equations are

$$
\begin{aligned}
\rho \ddot{u}_i &= \frac{\partial}{\partial x_j}\Big(a_{ijkh}\frac{\partial u_k}{\partial x_h}\Big) + \frac{\partial}{\partial x_j}(b_{ij}\nu) + \frac{\partial}{\partial x_j}(d_{ij}\omega) + \rho f_i , \\
\kappa_1 \ddot{\nu} &= \frac{\partial}{\partial x_i}\Big(\alpha_{ij}\frac{\partial \nu}{\partial x_j}\Big) + \frac{\partial}{\partial x_i}\Big(\beta_{ij}\frac{\partial \omega}{\partial x_j}\Big) - b_{ij}\frac{\partial u_i}{\partial x_j} - \alpha_1 \nu - \alpha_3 \omega + \rho \ell_1 , \\
\kappa_2 \ddot{\omega} &= \frac{\partial}{\partial x_i}\Big(\gamma_{ij}\frac{\partial \omega}{\partial x_j}\Big) + \frac{\partial}{\partial x_i}\Big(\beta_{ij}\frac{\partial \nu}{\partial x_j}\Big) - d_{ij}\frac{\partial u_i}{\partial x_j} - \alpha_2 \omega - \alpha_3 \nu + \rho \ell_2 .
\end{aligned}
\tag{4.5}
$$

In these equations the elasticities satisfy (4.2) while the tensors $b_{ij}, d_{ij}, \alpha_{ij}, \beta_{ij}$ and γ_{ij} are symmetric.

Before embarking on an analysis of the solution behaviour to equations (4.1), (4.3), (4.4) and (4.5) via an investigation with acceleration waves and with logarithmic convexity, it is pertinent to remark on differences to the structure of the equations. We observe that equations (4.1) are essentially a combination of a hyperbolic equation and a parabolic one, whereas equations (4.3) is a coupled system of two hyperbolic equations. Thus, even though equations (4.1) and (4.3) ostensibly model the same physical behaviour they are mathematically very different objects. Indeed, we find that the physical behaviour predicted is different. When we examine the equations for dual porosity elasticity, namely (4.4) and (4.5) we see that (4.4) consists of a hyperbolic equation and what are effectively two coupled parabolic equations. Equations (4.5) on the other hand are essentially three coupled hyperbolic equations. This suggests that their mathematical treatment and physical predictions will be substantially different, and this is precisely what we find.

In the interests of clarity we include a table of the physical variables and their meaning and these are provided in table 4.1.

Table 4.1 Physical variables and their definitions

	Symbol	Physical Meaning
One porosity theory	u_i	elastic displacement
	p	pressure in the pores
One void theory	u_i	elastic displacement
	ν	void volume distribution function
Double porosity theory	u_i	elastic displacement
	p	pressure in the macro pores
	q	pressure in the micro pores
Double voids theory	u_i	elastic displacement
	ν	void volume distribution function in the macro pores
	ω	void volume distribution function in the micro pores

4.2 Linear Acceleration Waves

4.2.1 One Porosity Theory

We introduce the notion of an acceleration wave in the context of an elastic body with one porosity field. Thus, the governing equations are (4.1). An acceleration wave for the system (4.1) is a two-dimensional surface, $\mathscr{S}$, in $\mathbb{R}^3$, such that $u_i(\mathbf{x},t)$ and $p(\mathbf{x},t)$ are C^1 everywhere but across $\mathscr{S}$, $\ddot{u}_i, \dot{u}_{i,j}, u_{i,jk}$, $\ddot{p}, \dot{p}_{,i}, p_{,ij}$ and their higher derivatives possess a finite discontinuity. We assume f_i and s_1 are C^∞ or without loss of generality we set them equal to zero. A key term in the theory of acceleration waves is the idea of the jump of a function f across $\mathscr{S}$. This is denoted by $[f]$ and is defined as

$$[f] = f^- - f^+, \tag{4.6}$$

where

$$f^+ = \lim_{\mathbf{x}\to\mathscr{S}} f(\mathbf{x},t) \qquad \text{from the right,}$$

and

$$f^- = \lim_{\mathbf{x}\to\mathscr{S}} f(\mathbf{x},t) \qquad \text{from the left.}$$

The amplitudes $a_i(t)$ and $P(t)$ of the acceleration wave are defined as

$$a_i(t) = [\ddot{u}_i] \qquad \text{and} \qquad P(t) = [\ddot{p}]. \tag{4.7}$$

To perform a mathematical analysis for an acceleration wave one requires compatibility conditions relating variables across the wave $\mathscr{S}$. These may be found in great detail in Truesdell & Toupin [221], pp. 491–525, although the conditions necessary for the computations performed here are conveniently found in Straughan [202], pp. 297–374, or Straughan [203], pp. 100–136.

We commence by taking the jumps of equations $(4.1)_1$ and $(4.1)_2$. Recalling the differentiability properties of u_i and p the result of this procedure leads to

$$\rho[\ddot{u}_i] = a_{ijkh}[u_{k,hj}], \tag{4.8}$$

and

$$k_{ij}[p_{,ji}] = \beta_{ij}[\dot{u}_{i,j}]. \tag{4.9}$$

Let U_N denote the wavespeed of $\mathscr{S}$ and then one has the Hadamard relation

$$\frac{\delta}{\delta t}[f] = [\dot{f}] + U_N[f_{,i}n_i] \tag{4.10}$$

where $\delta/\delta t$ is the displacement derivative, i.e. the rate of change of a quantity as seen by an observer on the wave. In equation (4.10) n_i is the unit outward normal to $\mathscr{S}$. In addition, we require the compatibility conditions

$$[f_{,i}] = n_i[f_{,j}n_j] \qquad \text{and} \qquad [f_{,ij}] = n_i n_j[f_{,rs}n_r n_s]. \tag{4.11}$$

Recollecting the differentiability of u_i one may use the Hadamard relation (4.10) to find that

$$[\ddot{u}_i] = -U_N[n_a\dot{u}_{i,a}] = U_N^2[n_a n_b u_{i,ab}]. \tag{4.12}$$

The compatibility condition allows one to obtain

$$[u_{k,hj}] = n_h n_j [u_{k,rs} n_r n_s]. \tag{4.13}$$

We also apply the Hadamard condition to p to obtain

$$[\ddot{p}] = U_N^2[n_i n_j p_{,ij}]. \tag{4.14}$$

The next step involves using (4.12) and (4.13) in equation (4.8) to obtain

$$(\rho U_N^2 \delta_{ij} - Q_{ij})a_j = 0. \tag{4.15}$$

In this equation Q_{ij} is the acoustic tensor given by

$$Q_{ij} = a_{irjs} n_r n_s. \tag{4.16}$$

Suppose we may write $a_i = a(t)\lambda_i$ then from (4.15) we obtain the fact that an acceleration wave will propagate when λ is an eigenvector of $\mathbf{Q}$. If we know the direction of λ then the wavespeed is given by (4.15) as

$$U_N^2 = \frac{Q_{ij}\lambda_i\lambda_j}{\rho}. \tag{4.17}$$

Thus, this theory predicts two waves, each with speed U_N, but moving in opposite directions, and these are predominantly associated to the elastic behaviour of the body. A particular type of acceleration wave is found when $\lambda_i = n_i$ and then one says it is a longitudinal wave.

A separate wavespeed is not found in connection with the fluid pressure via the amplitude P. Instead, employing the compatibility relations the jump of (4.9) leads to

$$\frac{k_{ij}n_i n_j}{U_N^2} P = -\frac{\beta_{ij}n_j}{U_N} a_i. \tag{4.18}$$

Thus, the quantity P is directly connected to the amplitude a_i. If we know a_i then P follows from (4.18).

One may determine the amplitude a_i exactly. However, for purposes of illustration we restrict the calculation of the amplitude equation to a plane wave in three-dimensions and this is essentially equivalent to dealing with the amplitude equation in one-dimension. Thus, we now deal with a one-dimensional acceleration wave moving along the $x-$axis.

Equations (4.1) in one space dimension are

$$\begin{aligned} \rho\ddot{u} &= Au_{xx} - \beta p_x, \\ \alpha\dot{p} &= kp_{xx} - \beta\dot{u}_x. \end{aligned} \tag{4.19}$$

Performing a calculation like that leading to (4.17), but in one space dimension, one may show that the one-dimensional wavespeed V is given by

$$V^2 = \frac{A}{\rho}.$$

This follows from the jump relation arising from $(4.19)_1$, i.e.

$$\rho[\ddot{u}] = A[u_{xx}].$$

The jump relation from $(4.19)_2$ gives

$$\frac{kP}{V^2} = -\frac{\beta a}{V}, \tag{4.20}$$

where now $P = [\ddot{p}]$ and $a = [\ddot{u}]$.

To determine the amplitude a (or P) we differentiate $(4.19)_1$ with respect to x and take the jump of the resulting equation to see that

$$\rho[\ddot{u}_x] = A[u_{xxx}] - \beta[p_{xx}]. \tag{4.21}$$

Upon using the Hadamard relation one obtains

$$[\ddot{u}] = -V[\dot{u}_x] = V^2[u_{xx}], \tag{4.22}$$

and also

$$\frac{\delta}{\delta t}[\dot{u}_x] = [\ddot{u}_x] + V[\dot{u}_{xx}] \qquad \text{and} \qquad \frac{\delta}{\delta t}[u_{xx}] = [\dot{u}_{xx}] + V[u_{xxx}]. \tag{4.23}$$

Next, from both expressions in (4.23) one shows

$$[\ddot{u}_x] = -V\left(-V[u_{xxx}] + \frac{\delta}{\delta t}\frac{a}{V^2}\right) - \frac{\delta}{\delta t}\left(\frac{a}{V}\right)$$

and since $V = \sqrt{A/\rho}$ which is constant,

$$[\ddot{u}_x] = V^2[u_{xxx}] - \frac{2}{V}\frac{\delta a}{\delta t}. \tag{4.24}$$

The term $[\ddot{u}_x]$ is substituted from (4.24) in (4.21) to find

$$-\frac{2\rho}{V}\frac{\delta a}{\delta t} + \rho V^2[u_{xxx}] = A[u_{xxx}] - \frac{\beta P}{V^2}.$$

Now since $V^2 = A/\rho$ the $[u_{xxx}]$ term disappears and then upon employing (4.20) to substitute for P we arrive at the amplitude equation,

$$\frac{\delta a}{\delta t} = -\frac{\beta^2}{2\rho k V^2}a. \tag{4.25}$$

This equation integrates to see that

$$a(t) = \exp\left(\frac{-\beta^2 t}{2\rho k V^2}\right) a(0).$$

Hence, the amplitude a decreases exponentially. From (4.20) one deduces P also has exponential decay in time.

4.2.2 One Void Theory

We commence with the definition of an acceleration wave in the case of an elastic body with a single distribution of voids. An acceleration wave for the system (4.3) is a two-dimensional surface, $\mathscr{S}$, in $\mathbb{R}^3$, such that $u_i(\mathbf{x},t)$ and $\nu(\mathbf{x},t)$ are C^1 everywhere but across $\mathscr{S}$, $\ddot{u}_i, \dot{u}_{i,j}, u_{i,jk}$, $\ddot{\nu}, \dot{\nu}_{,i}, \nu_{,ij}$ and their higher derivatives possess a finite discontinuity. The functions f_i and ℓ are C^∞ or we may simply set them equal to zero. The amplitudes $a_i(t)$ and $B(t)$ of the acceleration wave are defined as

$$a_i(t) = [\ddot{u}_i] \qquad \text{and} \qquad B(t) = [\ddot{\nu}]. \tag{4.26}$$

We employ relations (4.10) - (4.14) of section 4.2.1. Upon utilizing the regularity assumptions of the acceleration wave we take the jumps of equations (4.3) across $\mathscr{S}$ to find

$$\rho[\ddot{u}_i] = a_{ijkh}[u_{k,hj}] \tag{4.27}$$

and

$$\kappa[\ddot{\nu}] = \alpha_{ij}[\nu_{,ij}]. \tag{4.28}$$

Equation (4.27) is handled exactly as in section 4.2.1. We again find that

$$(\rho U_N^2 \delta_{ij} - Q_{ij})a_j = 0. \tag{4.29}$$

To handle equation (4.28) we note that from the Hadamard and compatibility relations we see that

$$[\nu_{,ij}] = n_i n_j [\nu_{,rs} n_r n_s], \qquad [\ddot{\nu}] = U_N^2 [\nu_{,rs} n_r n_s]$$

and so from (4.28) one has

$$(\kappa_1 U_N^2 - Q_1)B = 0, \tag{4.30}$$

where Q_1 is a void-like acoustic term given by

$$Q_1 = \alpha_{ij} n_i n_j. \tag{4.31}$$

Thus, for the one void material we see that there are *two* right and left moving waves given by

$$U_{N(1)}^2 = \frac{Q_{ij}n_i n_j}{\rho} \qquad \text{and} \qquad U_{N(2)}^2 = \frac{Q_1}{\kappa_1}. \tag{4.32}$$

In general, one might expect $U_{N(1)} \neq U_{N(2)}$ and then we can argue that there are two separate waves, one associated with the elastic nature of the body whereas the other is directly connected to the voids distribution.

To derive the behaviour of the amplitudes we again restrict attention to a one-dimensional wave. Equations (4.3) in one-dimension may be written

$$\begin{aligned} \rho \ddot{u} &= A u_{xx} + \beta \nu_x, \\ \kappa_1 \ddot{\nu} &= \alpha \nu_{xx} - \beta u_x - \xi \nu. \end{aligned} \tag{4.33}$$

The wavespeed equations are then

$$(\rho V^2 - A)a = 0,$$

and

$$(\kappa_1 V^2 - \alpha)B = 0,$$

where V is the one-dimensional wavespeed. We may write $V_1^2 = A/\rho$ and $V_2^2 = \alpha/\kappa_1$. To find the wave amplitudes we differentiate (4.33) with respect to x and take the jumps across $\mathscr{S}$ to find

$$\begin{aligned} \rho [\ddot{u}_x] &= A[u_{xxx}] + \beta [\nu_{xx}], \\ \kappa_1 [\ddot{\nu}_x] &= \alpha [\nu_{xxx}] - \beta [u_{xx}]. \end{aligned} \tag{4.34}$$

Next, employ (4.24) in $(4.34)_1$ and also use the one-dimensional equivalent of (4.28) to obtain

$$\frac{\delta a}{\delta t} = -\frac{\beta \kappa_1 V}{2\rho\alpha} B, \tag{4.35}$$

and with a similar analysis we derive from (4.34)

$$\frac{\delta B}{\delta t} = \frac{\rho \beta V}{2\kappa_1 A} a. \tag{4.36}$$

It appears at this point that one may argue from (4.32) that there are two disconnected waves with speeds V_1 and V_2 and, in general, these wavespeeds will be different. One point of view is then that on the wave with speed V_1 only $\ddot{u}$ has a discontinuity with $\ddot{\nu}$ remaining continuous and on the wave with speed V_2 only $\ddot{\nu}$ is discontinuous while $\ddot{u}$ stays continuous. In this case one will have two distinct waves and from (4.35) one has $\delta a/\delta t = 0$ on wave one while on wave two, $\delta B/\delta t = 0$. Thus, there are two different waves, moving with different speeds but the respective wave amplitude stays constant on each wave. In the exceptional case where $V_1 = V_2$ then one may eliminate a or B from (4.35) and (4.36) to find, for example,

$$\frac{\delta^2 a}{\delta t^2} + \frac{\beta^2 V^2}{4\alpha A} a = 0. \tag{4.37}$$

Thus, in this case the amplitude oscillates since a satisfies a simple harmonic motion equation.

To sum up the comparison situation we see that for the single porosity model of section 4.2.1 there is only one acceleration wave and the wave amplitude decays exponentially. For the single void distribution case there are, in general, two distinct acceleration waves and the amplitude of each appears to remain constant. Thus, even for the one porosity or one void case the solutions to the two models behave very differently. We stress that we have only analysed the situation from a *linear* theory at this stage. The nonlinear theory has the potential to behave in a different manner, see chapters 8 and 9, although then the key part in the momentum equation is what the Piola-Kirchoff stress tensor depends on, i.e. what the internal energy function depends on.

4.2.3 Double Porosity Theory

We only give brief details in this case since the method is similar to that in section 4.2.1 and full details may be found in Straughan [205].

The relevant equations are (4.4) and we define an acceleration wave for these equations to be a two-dimensional surface, $\mathscr{S}$, in $\mathbb{R}^3$, such that u_i, p and q are C^1 everywhere but across $\mathscr{S}$, $\ddot{u}_i, \dot{u}_{i,j}, u_{i,jk}, \ddot{p}, \dot{p}_{,i}, p_{,ij}, \ddot{q}, \dot{q}_{,i}, q_{,ij}$ and their higher derivatives possess a finite discontinuity. Again, we assume f_i, s_1 and s_2 are C^∞ or zero.

The analysis begins by taking the jumps of equations (4.4) and then we employ the Hadamard relation and compatibility conditions. Straughan [205] shows that the amplitudes $a(t) = [\ddot{u}_i]$, $P(t) = [\ddot{p}]$ and $Q(t) = [\ddot{q}]$ satisfy the equations

$$\begin{aligned}
(\rho U_N^2 \delta_{ij} - Q_{ij})a_j &= 0, \\
k_{ij}n_i n_j P &= -\frac{\beta_{ij}n_j}{U_N}\, a_i\,, \\
m_{ij}n_i n_j Q &= -\frac{\gamma_{ij}n_j}{U_N}\, a_i\,.
\end{aligned} \tag{4.38}$$

Again, an acceleration wave with $a_i = a(t)\lambda_i$ will propagate when λ is an eigenvalue of $\mathbf{Q}$ and the wavespeed is given by $U_N^2 = Q_{ij}\lambda_i\lambda_j/\rho$. In this case there are two waves, which move in both right and left directions with speed U_N, and we concentrate on one wave, namely that moving to the right. Equations $(4.38)_{2,3}$ serve to yield the amplitudes P and Q once a_i is known.

By adopting an analysis similar to that of section 4.2.1 Straughan [205] shows that the one-dimensional amplitude a satisfies the equation

$$\frac{\delta a}{\delta t} = -\frac{1}{2\rho V^2}\left(\frac{\beta^2}{k} + \frac{\gamma^2}{m}\right)a$$

where β, γ, k and m are the one-dimensional equivalents of $\beta_{ij}, \gamma_{ij}, k_{ij}$ and m_{ij}. Thus, a, P and Q again decay exponentially.

4.2.4 Double Voids Theory

For this section the appropriate equations are (4.5) and an acceleration wave is defined to be a two-dimensional surface, $\mathscr{S}$, in $\mathbb{R}^3$, such that u_i, ν and ω are C^1 everywhere but across $\mathscr{S}$, $\ddot{u}_i, \dot{u}_{i,j}, u_{i,jk}$, $\ddot{\nu}$, $\dot{\nu}_{,i}$, $\nu_{,ij}$, $\ddot{\omega}, \dot{\omega}_{,i}, \omega_{,ij}$ and their higher derivatives possess a finite discontinuity. Again, the source terms f_i, ℓ_1 and ℓ_2 are C^∞ or zero.

The relevant equations we shall employ are (3.69) but in an isothermal situation. Thus, the equations to be analysed here are

$$\begin{aligned}\rho_0\ddot{u}_i &= (a_{ijkh}u_{k,h})_{,j} + (b_{ij}\nu)_{,j} + (d_{ij}\omega)_{,j} + \rho_0 f_i\,,\\ \kappa_1\ddot{\nu} &= (\alpha_{ij}\nu_{,j})_{,i} + (\beta_{ij}\omega_{,j})_{,i} - b_{ij}u_{i,j} - \alpha_1\nu - \alpha_3\omega + \rho_0\ell_1\,,\\ \kappa_2\ddot{\omega} &= (\gamma_{ij}\omega_{,j})_{,i} + (\beta_{ij}\nu_{,j})_{,i} - d_{ij}u_{i,j} - \alpha_2\omega - \alpha_3\nu + \rho_0\ell_2\,.\end{aligned} \tag{4.39}$$

Define the wave amplitudes now as $a_i = [\ddot{u}_i]$, $B = [\ddot{\nu}]$, $C = [\ddot{\omega}]$. The acceleration wave analysis commences by taking the jump of equation $(4.39)_1$ to find

$$\begin{aligned}\rho_0[\ddot{u}_i] &= a_{ijkh}[u_{k,hj}]\\ &= a_{ijkh}n_jn_h[u_{k,rs}n_rn_s]\\ &= a_{ijkh}n_jn_h\,\frac{a_k}{U_N^2}\end{aligned}$$

where the Hadamard relation and compatibility conditions have been employed. This leads to the equation

$$(\rho_0U_N^2\delta_{ij} - Q_{ij})a_j = 0. \tag{4.40}$$

In (4.40) Q_{ij} is the acoustic tensor. From the jumps of equations $(4.39)_{2,3}$ one derives

$$\begin{aligned}\kappa_1[\ddot{\nu}] &= \alpha_{ij}[\nu_{,ij}] + \beta_{ij}[\omega_{,ij}]\,,\\ \kappa_2[\ddot{\omega}] &= \gamma_{ij}[\omega_{,ij}] + \beta_{ij}[\nu_{,ij}]\,.\end{aligned} \tag{4.41}$$

By further use of the Hadamard relation and compatibility conditions these equations lead to

$$\begin{aligned}\kappa_1U_N^2B &= \alpha_{ij}n_in_jB + \beta_{ij}n_in_jC\,,\\ \kappa_2U_N^2C &= \gamma_{ij}n_in_jC + \beta_{ij}n_in_jB\,.\end{aligned} \tag{4.42}$$

Let $Q_1 = \alpha_{ij}n_in_j, Q_2 = \gamma_{ij}n_in_j$ and $Q_3 = \beta_{ij}n_in_j$. Then (4.42) may be rearranged as

$$\begin{pmatrix}\kappa_1U_N^2 - Q_1 & -Q_3\\ -Q_3 & \kappa_2U_N^2 - Q_2\end{pmatrix}\begin{pmatrix}B\\ C\end{pmatrix} = \begin{pmatrix}0\\ 0\end{pmatrix}.$$

For non-zero amplitudes we then need

$$(\kappa_1 U_N^2 - Q_1)(\kappa_2 U_N^2 - Q_2) - Q_3^2 = 0. \tag{4.43}$$

Define $U_M^2 = Q_1/\kappa_1$ and $U_m^2 = Q_2/\kappa_2$, then (4.43) may be rearranged as

$$U_N^4 - (U_M^2 + U_m^2)U_N^2 + U_M^2 U_m^2 - P^2 = 0, \tag{4.44}$$

where $P^2 = Q_3^2/\kappa_1\kappa_2$. This equation has solution,

$$U_N^2 = \frac{1}{2}\{U_M^2 + U_m^2 \pm \sqrt{(U_M^2 + U_m^2)^2 - 4(U_M^2 U_m^2 - P^2)}\}, \tag{4.45}$$

$$= \frac{1}{2}\{U_M^2 + U_m^2 \pm \sqrt{(U_M^2 - U_m^2)^2 + 4P^2}\}. \tag{4.46}$$

If $U_M^2 U_m^2 - P^2 > 0$ then we see from (4.45) that there are two positive solutions. This leads to two sets of waves (each set has a right going and a left going wave). Let these solutions be $U_1^2 \geq U_2^2$ where U_1^2 assumes the positive sign in (4.46) and where U_2^2 assumes the negative sign. We may then show

$$2(U_M^2 + U_m^2) \geq U_1^2 \geq U_M^2 + U_m^2 \geq U_2^2 > 0.$$

Thus, from (4.40) and (4.46) we see that there are three waves. The elastic one has wavespeed $U_N = \sqrt{Q_{ij}\lambda_i\lambda_j/\rho_0}$ where λ_i is an eigenvector of $\mathbf{Q}$. There is also a coupled pair of waves with speeds U_1 and U_2. The elastic wave is disconnected to the two coupled voids waves. This appears to suggest the elastic wave moves with $\ddot{u}_i$ discontinuous but $\ddot{v}$ and $\ddot{\omega}$ continuous, whereas on the voids waves, $\ddot{v}$ and $\ddot{\omega}$ are discontinuous with $\ddot{u}_i$ continuous.

We briefly describe the calculation of the amplitudes for a one-dimensional wave. The one-dimensional equations from (4.39) are, with zero body force and voids supplies, with the density now denoted by ρ,

$$\begin{aligned} \rho\ddot{u} &= Au_{xx} + bv_x + d\omega_x, \\ \kappa_1\ddot{v} &= \alpha v_{xx} + \beta\omega_{xx} - bu_x - \alpha_1 v - \alpha_3\omega, \\ \kappa_2\ddot{\omega} &= \gamma\omega_{xx} + \beta v_{xx} - du_x - \alpha_3 v - \alpha_2\omega. \end{aligned} \tag{4.47}$$

We take the x derivative of (4.47) and then take the jumps to obtain

$$\begin{aligned} \rho[\ddot{u}_x] &= A[u_{xxx}] + b[v_{xx}] + d[\omega_{xx}], \\ \kappa_1[\ddot{v}_x] &= \alpha[v_{xxx}] + \beta[\omega_{xxx}] - b[u_{xx}], \\ \kappa_2[\ddot{\omega}_x] &= \gamma[\omega_{xxx}] + \beta[v_{xxx}] - d[u_{xx}]. \end{aligned} \tag{4.48}$$

When we are in the situation where the elastic wave is disconnected then the last two terms in $(4.48)_1$ are not present. One may then employ (4.24) in $(4.48)_1$ to see that

$$-\frac{2\rho}{V}\frac{\delta a}{\delta t} + \rho V^2[u_{xxx}] = A[u_{xxx}].$$

The one-dimensional wavespeed V of the elastic wave is $V=\sqrt{A/\rho}$ and then one sees $\delta a/\delta t=0$. Hence, the amplitude on the elastic wave is constant. For the equations $(4.48)_2$ the $[u_{xx}]$ terms disappear and then one employs (4.24) with u replaced by ν or ω. In this case one obtains

$$\begin{aligned}-\frac{2\kappa_1}{V}\frac{\delta B}{\delta t}+\kappa_1V^2[\nu_{xxx}]&=\alpha[\nu_{xxx}]+\beta[\omega_{xxx}],\\ -\frac{2\kappa_2}{V}\frac{\delta C}{\delta t}+\kappa_2V^2[\omega_{xxx}]&=\gamma[\omega_{xxx}]+\beta[\nu_{xxx}].\end{aligned}\tag{4.49}$$

The one-dimensional wavespeed equations are

$$(\kappa_1V^2-\alpha)B=\beta C\qquad\text{and}\qquad(\kappa_2V^2-\gamma)C=\beta B$$

and so

$$(\kappa_1V^2-\alpha)(\kappa_2V^2-\gamma)-\beta^2=0.\tag{4.50}$$

One forms $(4.49)_1+\lambda(4.49)_2$ for a number λ to be determined. This leads to

$$\begin{aligned}-\left\{\frac{2\kappa_1}{V}+\frac{2\kappa_2\lambda}{V}\left(\frac{\kappa_1V^2-\alpha}{\beta}\right)\right\}\frac{\delta B}{\delta t}+(\kappa_1V^2-\alpha-\lambda\beta)[\nu_{xxx}]\\ +(\lambda\kappa_2V^2-\lambda\gamma-\beta)[\omega_{xxx}]=0.\end{aligned}\tag{4.51}$$

Select now $\lambda=(\kappa_1V^2-\alpha)/\beta$ and note that from (4.50) one has also $\lambda=\beta/(\kappa_2V^2-\gamma)$. Then from $(4.49)_2$ we find $\delta B/\delta t=0$ and so B and then C have constant amplitudes.

Thus, for comparison, again the double porosity and double void theories lead to different results when handling acceleration waves. The double porosity case has only one wave whose amplitude decays exponentially. The double voids case yields three waves all of which have constant amplitudes.

4.3 Uniqueness

In the next sections we consider differences between the proofs of uniqueness for the four models for which we have made a comparison by using acceleration waves.

4.3.1 Uniqueness, Single Porosity

We begin with the one porosity model of section 4.1.1. We suppose equations (4.1) are defined on a bounded domain Ω with boundary Γ. The boundary conditions are

$$u_i=u_i^B(\mathbf{x},t),\qquad p=p^B(\mathbf{x},t),\qquad\text{on }\gamma\times(0,T),\tag{4.52}$$

where u_i^B and p^B are prescribed functions. The solution satisfies the initial conditions

$$u_i(\mathbf{x},0) = v_i(\mathbf{x}), \qquad \dot{u}_i(\mathbf{x},0) = w_i(\mathbf{x}), \qquad p(\mathbf{x},0) = f(\mathbf{x}), \tag{4.53}$$

where v_i, w_i and f_i are given functions. The assumptions made on the coefficients are that

$$a_{ijkh} = a_{khij} = a_{jikh}\,, \tag{4.54}$$

where b_{ij} and k_{ij} are symmetric and

$$k_{ij}\xi_i\xi_j \geq 0 \qquad \forall \xi_i\,. \tag{4.55}$$

To consider the uniqueness question we let (u_i^1, p^1) and (u_i^2, p^2) be two solutions to (4.1) with both solutions satisfying these equations for the same body force and source functions f_i, s_1 and both solutions satisfying (4.52) and (4.53) for the same data functions u_i^B, p^B, v_i, w_i and f. Let (u_i, p) be the difference solution defined by

$$u_i = u_i^1 - u_i^2, \qquad p = p^1 - p^2.$$

One may then see that the difference solution satisfies the equations

$$\begin{aligned} \rho\ddot{u}_i &= (a_{ijkh}u_{k,h})_{,j} - (\beta_{ij}p)_{,j}\,, \\ \alpha\dot{p} &= (k_{ij}p_{,j})_{,i} - \beta_{ij}\dot{u}_{i,j}\,, \end{aligned} \tag{4.56}$$

and the boundary and initial conditions,

$$u_i = 0, \quad p = 0, \qquad \text{on } \Gamma \times (0,T), \tag{4.57}$$

and

$$u_i(\mathbf{x},0) = 0, \quad \dot{u}_i(\mathbf{x},0) = 0, \quad p(\mathbf{x},0) = 0, \qquad \mathbf{x} \in \Omega. \tag{4.58}$$

To establish uniqueness we employ a logarithmic convexity method. Define the function $F(t)$ by

$$F(t) = <\rho u_i u_i> + \int_0^t < k_{ij}\eta_{,i}\eta_{,j} > ds, \tag{4.59}$$

where $< \cdot >$ denotes integration over Ω, and

$$\eta(\mathbf{x},t) = \int_0^t p(\mathbf{x},s)ds. \tag{4.60}$$

Upon differentiation one finds, using the inital conditions (4.58),

$$F' = 2 < \rho u_i \dot{u}_i > +2\int_0^t < k_{ij}\eta_{,i}p_{,j} > ds. \tag{4.61}$$

Then, upon a further differentiation we see that

$$F'' = 2 < \rho \dot{u}_i \dot{u}_i > +2 < \rho u_i \ddot{u}_i > +2 < k_{ij} \eta_{,i} p_{,j} > . \tag{4.62}$$

To proceed we shall need to integrate equation $(4.56)_2$ with respect to time and employ the initial conditions to obtain the equation

$$\alpha p = (k_{ij} \eta_{,j})_{,i} - \beta_{ij} u_{i,j} . \tag{4.63}$$

This equation is multiplied by p and integrated over Ω, and then integrating by parts and using the boundary conditions one obtains

$$- < \alpha p^2 > - < \beta_{ij} u_{i,j} p > = < k_{ij} \eta_{,j} p_{,i} > . \tag{4.64}$$

Next, add and subtract $2 < \rho \dot{u}_i \dot{u}_i >$ in (4.62), and substitute for $\rho \ddot{u}_i$ using equation (4.56) to find

$$\begin{aligned} F'' =& 4 < \rho \dot{u}_i \dot{u}_i > -2 < \rho \dot{u}_i \dot{u}_i > +2 < u_i (a_{ijkh} u_{k,h})_{,j} > \\ & -2 < u_i (\beta_{ij} p)_{,j} > +2 < k_{ij} \eta_{,j} p_{,i} > . \end{aligned} \tag{4.65}$$

Integrate by parts using the boundary conditions and substitute for the last term from (4.64) to find

$$\begin{aligned} F'' =& 4 < \rho \dot{u}_i \dot{u}_i > -2 < \rho \dot{u}_i \dot{u}_i > \\ & -2 < a_{ijkh} u_{k,h} u_{i,j} > -2 < \alpha p^2 > . \end{aligned} \tag{4.66}$$

One derives an energy equation by multiplying $(4.56)_1$ by $\dot{u}_i$ and integrating over Ω, then by multiplying $(4.56)_2$ by p and integrating over Ω. The result is integrated in time to find

$$E(t) + \int_0^t < k_{ij} p_{,i} p_{,j} > ds = E(0) , \tag{4.67}$$

where $E(t)$ is the energy function

$$E(t) = \frac{1}{2} < \rho \dot{u}_i \dot{u}_i > + \frac{1}{2} < a_{ijkh} u_{k,h} u_{i,j} > + \frac{1}{2} < \alpha p^2 > . \tag{4.68}$$

For the uniqueness problem the initial conditions are (4.58) and thus $E(0) = 0$. Thus, from equation (4.66) one shows

$$\begin{aligned} F'' =& 4 < \rho \dot{u}_i \dot{u}_i > -4E(t) , \\ =& 4 < \rho \dot{u}_i \dot{u}_i > +4 \int_0^t < k_{ij} p_{,i} p_{,j} > ds. \end{aligned} \tag{4.69}$$

Upon using equation (4.59), (4.61) and (4.69) one shows

$$FF'' - (F')^2 = 4S^2 \geq 0, \tag{4.70}$$

where S^2 is given by

$$\begin{aligned} S^2 = & \left[< \rho u_i u_i > + \int_0^t < k_{ij}\eta_{,j}\eta_{,i} > ds \right] \\ & \times \left[< \rho \dot{u}_i \dot{u}_i > + \int_0^t < k_{ij} p_{,i} p_{,j} > ds \right] \\ & - \left[< \rho u_i \dot{u}_i > + \int_0^t < k_{ij}\eta_{,j} p_{,i} > ds \right]^2 . \end{aligned} \tag{4.71}$$

Inequality (4.70) demonstrates $\log F$ is a convex function of t. One then shows

$$F(t) \leq [F(0)]^{1-t/T} [F(T)]^{t/T}, \tag{4.72}$$

for an interval $(0,T)$, $T < \infty$, and since $F(0) = 0$ it follows that $F(t) \equiv 0$ and then $u_i \equiv 0$. Then from the energy equation (4.67) we deduce

$$0 \leq < \alpha p^2 > \leq 0$$

and so $p \equiv 0$ and uniqueness follows. (The argument is really more delicate than this since $F(0) = 0$ and taking the logarithm of this is not allowed. A rigorous proof is easily constructed by following the outline in section 1.3.6. The main focus in this chapter is on comparing porosity and voids theories and so we are concentrating on the differences in uniqueness proofs.)

4.3.2 Uniqueness, Single Voids

We now consider the analogous uniqueness question to that of section 4.3.1 but for the boundary-initial value problem for the one void distribution theory, namely, for equations (4.3). Similarly to section 4.3.1 we suppose equations (4.3) are defined on $\Omega \times (0,T]$, for some time $0 < T < \infty$. The boundary and initial conditions are

$$u_i = u_i^B(\mathbf{x},t), \qquad \nu = \nu^B(\mathbf{x},t), \qquad \text{on } \Gamma \times (0,T], \tag{4.73}$$

and

$$\begin{aligned} u_i(\mathbf{x},0) = v_i(\mathbf{x}), \qquad & \dot{u}_i(\mathbf{x},0) = w_i(\mathbf{x}), \\ \nu(\mathbf{x},0) = \phi(\mathbf{x}), \qquad & \dot{\nu}(\mathbf{x},0) = \hat{\phi}(\mathbf{x}). \end{aligned} \tag{4.74}$$

Let (u_i^1, ν^1) and (u_i^2, ν^2) be solutions to (4.3) together with (4.73) and (4.74), for the same functions f_i and ℓ, and for the same data functions $u_i^B, \nu^B, v_i, w_i, \phi$ and $\hat{\phi}$. Define the difference variables by

$$u_i = u_i^1 - u_i^2, \qquad \nu = \nu^1 - \nu^2,$$

and then (u_i, ν) satsifies the boundary-initial value problem

$$\begin{aligned} \rho \ddot{u}_i &= (a_{ijkh} u_{k,h})_{,j} + (b_{ij}\nu)_{,j}, \\ \kappa_1 \ddot{\nu} &= (\alpha_{ij}\nu_{,j})_{,i} - b_{ij}u_{i,j} - \xi\nu, \end{aligned} \tag{4.75}$$

with

$$\begin{aligned} &u_i(\mathbf{x},t)=0, \qquad \nu(\mathbf{x},t)=0, \qquad \text{on } \Gamma\times(0,T], \\ &u_i(\mathbf{x},0)=0, \quad \dot{u}_i(\mathbf{x},0)=0, \quad \nu(\mathbf{x},0)=0, \quad \dot{\nu}(\mathbf{x},0)=0, \qquad \mathbf{x}\in\Omega . \end{aligned} \tag{4.76}$$

We assume that the coefficients a_{ijkh} satisfy the symmetries (4.54), α_{ij} and β_{ij} are symmetric, $\xi \geq 0$, and

$$\alpha_{ij}\xi_i\xi_j \geq 0, \qquad \forall \xi_i . \tag{4.77}$$

To establish uniqueness for the boundary-initial value problem under consideration we must show $u_i \equiv 0, \nu \equiv 0$, and again we employ a logarithmic convexity method, but now the function F has form

$$F(t) =< \rho u_i u_i > + < \kappa_1 \nu^2 > . \tag{4.78}$$

We immediately note that the second term in F of (4.78) is of an entirely different structure to that of (4.59).

The uniqueness proof begins by differentiating F to find

$$F'(t) = 2 < \rho u_i \dot{u}_i > +2 < \kappa_1 \nu\dot{\nu} >, \tag{4.79}$$

and further

$$\begin{aligned} F''(t) =&2 < \rho \dot{u}_i \dot{u}_i > +2 < \kappa_1 \dot{\nu}^2 > +2 < \rho u_i \ddot{u}_i > +2 < \kappa_1 \nu\ddot{\nu} >, \\ =&4 < \rho \dot{u}_i \dot{u}_i > +4 < \kappa_1 \dot{\nu}^2 > -2 < \rho \dot{u}_i \dot{u}_i > -2 < \kappa_1 \dot{\nu}^2 > \\ &-2 < a_{ijkh} u_{i,j} u_{k,h} > -4 < b_{ij} \nu u_{i,j} > \\ &-2 < \alpha_{ij} \nu_{,i} \nu_{,j} > -2 < \xi \nu^2 >, \end{aligned} \tag{4.80}$$

where we have substituted for $\rho\ddot{u}_i$, $\kappa_1\ddot{\nu}$ and integrated by parts.

An energy equation is formed by multiplying $(4.75)_1$ by $\dot{u}_i$, $(4.75)_2$ by $\dot{\nu}$, adding and integrating over Ω to find after integration by parts and an integration in time,

$$E(t) = E(0), \tag{4.81}$$

where

$$\begin{aligned} E(t) =&\frac{1}{2} < \rho \dot{u}_i \dot{u}_i > + \frac{1}{2} < a_{ijkh} u_{i,j} u_{k,h} > + \frac{1}{2} < \kappa_1 \dot{\nu}^2 > \\ &+ < b_{ij} \nu u_{i,j} > + \frac{1}{2} < \alpha_{ij} \nu_{,i} \nu_{,j} > + \frac{1}{2} < \xi \nu^2 > . \end{aligned} \tag{4.82}$$

Next, use (4.82) to substitute for the last six terms in (4.82) and then employ (4.81) to see that

$$\begin{aligned} F'' =&4 < \rho \dot{u}_i \dot{u}_i > +4 < \kappa_1 \dot{\nu}^2 > -4E(t), \\ =&4 < \rho \dot{u}_i \dot{u}_i > +4 < \kappa_1 \dot{\nu}^2 > -4E(0). \end{aligned} \tag{4.83}$$

Since for the uniqueness question $E(0) = 0$ it follows that

$$F'' = 4 < \rho \dot{u}_i \dot{u}_i > + 4 < \kappa_1 \dot{\nu}^2 > . \tag{4.84}$$

Now form $FF'' - (F')^2$ to find

$$FF'' - (F')^2 = 4S^2 \geq 0, \tag{4.85}$$

where now S^2 is given by

$$\begin{aligned} S^2 =& \big[< \rho u_i u_i > + < \kappa_1 \nu^2 >\big] \big[< \rho \dot{u}_i \dot{u}_i > + < \kappa_1 \dot{\nu}^2 >\big] \\ & - \big[< \rho u_i \dot{u}_i > + < \kappa_1 \nu \dot{\nu} >\big]^2 . \end{aligned} \tag{4.86}$$

From inequality (4.85) it follows that

$$(\log F)'' \geq 0 .$$

This leads to $F \equiv 0$, following an argument analogous to those in section 1.3.6, and then $u_i \equiv 0$, $\nu \equiv 0$ and so uniqueness holds.

From the comparison point of view sections 4.3.1 and 4.3.3 are very different. The functions F in both sections are different and the proofs differ strongly. Thus, we conclude that the porosity theory and the voids theory are very different mathematical objects.

4.3.3 Uniqueness, Double Porosity

For the double porosity elastic body case the equations are (4.4). As in sections 4.3.1 and 4.3.2 we consider two solutions (u_i^1, p^1, q^1) and (u_i^2, p^2, q^2) which satisfy (4.4) for the same data and forcing functions and then we write the boundary-inital value problem for the difference solution $u_i = u_i^1 - u_i^2$, $p = p^1 - p^2$ and $q = q^1 - q^2$. The uniqueness question in this case was analysed by Straughan [204].

One shows that (u_i, p, q) satisfy the equations

$$\begin{aligned} \rho \ddot{u}_i &= (a_{ijkh} u_{k,h})_{,j} - (\beta_{ij} p)_{,j} - (\gamma_{ij} q)_{,j} , \\ \alpha \dot{p} &= (k_{ij} p_{,j})_{,i} - \gamma (p - q) - \beta_{ij} \dot{u}_{i,j} , \\ \beta \dot{q} &= (m_{ij} q_{,j})_{,i} + \gamma (p - q) - \gamma_{ij} \dot{u}_{i,j} , \end{aligned} \tag{4.87}$$

with the boundary conditions

$$u_i = 0, \qquad p = 0, \qquad q = 0, \qquad \text{on } \Gamma \times (0, T], \tag{4.88}$$

and the initial conditions

$$u_i(\mathbf{x}, 0) = 0, \quad \dot{u}_i(\mathbf{x}, 0) = 0, \quad p(\mathbf{x}, 0) = 0, \quad q(\mathbf{x}, 0) = 0, \qquad \mathbf{x} \in \Omega . \tag{4.89}$$

The elasticities satisfy the symmetries (4.54), $\gamma > 0$, $\beta_{ij}, \gamma_{ij}, k_{ij}, m_{ij}$ are symmetric and

$$k_{ij}\xi_i\xi_j \geq 0, \qquad m_{ij}\xi_i\xi_j \geq 0, \qquad \forall \xi_i. \tag{4.90}$$

The uniqueness proof involves the introduction of the variables η and ζ which are defined by

$$\eta(\mathbf{x},t) = \int_0^t p(\mathbf{x},s)ds \qquad \text{and} \qquad \zeta(\mathbf{x},t) = \int_0^t q(\mathbf{x},s)ds. \tag{4.91}$$

We shall need equations obtained by integrating $(4.87)_{2,3}$ in time, namely

$$\begin{aligned} \alpha p &= (k_{ij}\eta_{,j})_{,i} - \beta_{ij}u_{i,j} - \gamma(\eta - \zeta), \\ \beta q &= (m_{ij}\zeta_{,j})_{,i} - \gamma_{ij}u_{i,j} + \gamma(\eta - \zeta). \end{aligned} \tag{4.92}$$

By multiplying $(4.87)_{1,2,3}$ by $\dot{u}_i, p, q$ in turn, integrating over Ω then integrating in time, adding the results and some integration by parts one shows the energy

$$E(t) = \frac{1}{2} < \rho\dot{u}_i\dot{u}_i > + \frac{1}{2} < \alpha p^2 > + \frac{1}{2} < \beta q^2 > + \frac{1}{2} < a_{ijkh}u_{i,j}u_{k,h} > \tag{4.93}$$

satsifies the energy equation

$$\begin{aligned} E(t) + \int_0^t < k_{ij}p_{,i}p_{,j} > ds + \int_0^t < m_{ij}q_{,i}q_{,j} > ds \\ + \int_0^t < \gamma(p-q)^2 > ds = E(0). \end{aligned} \tag{4.94}$$

We again use a logarithmic convexity argument. The function F is now

$$\begin{aligned} F(t) = < \rho u_i u_i > + \int_0^t < k_{ij}\eta_{,i}\eta_{,j} > ds + \int_0^t < m_{ij}\zeta_{,i}\zeta_{,j} > ds \\ + \int_0^t < \gamma(\eta - \zeta)^2 > ds. \end{aligned} \tag{4.95}$$

Upon differentiation,

$$\begin{aligned} F' = 2 < \rho u_i\dot{u}_i > + 2\int_0^t < k_{ij}\eta_{,i}p_{,j} > ds + 2\int_0^t < m_{ij}\zeta_{,i}q_{,j} > ds \\ + 2\int_0^t < \gamma(\eta - \zeta)(p - q) > ds, \end{aligned} \tag{4.96}$$

where in obtaining the last three terms we use the facts that p and q are zero at time $t = 0$.

Next differentiate F' to find

$$\begin{aligned}F'' =&2 < \rho\dot{u}_i\dot{u}_i > +2 < \rho u_i\ddot{u}_i > +2 < k_{ij}\eta_{,i}p_{,j} >\\&+2 < m_{ij}\zeta_{,i}q_{,j} > +2 < \gamma(\eta-\zeta)(p-q) >\\=&4 < \rho\dot{u}_i\dot{u}_i > -2 < \rho\dot{u}_i\dot{u}_i > -2 < a_{ijkh}u_{i,j}u_{k,h} >\\&+2 < \beta_{ij}u_{i,j}p > +2 < \gamma_{ij}u_{i,j}q > +2 < k_{ij}\eta_{,i}p_{,j} >\\&+2 < m_{ij}\zeta_{,i}q_{,j} > +2 < \gamma(\eta-\zeta)(p-q) >,\end{aligned}\tag{4.97}$$

where we have substituted for $\rho\ddot{u}_i$ from equation $(4.87)_1$ and integrated by parts.

The next step is to multiply $(4.92)_{1,2}$ by p and q, respectively, and integrate over Ω to find with some integration by parts,

$$\begin{aligned}&< \alpha p^2 > + < k_{ij}\eta_{,i}p_{,j} > + < \beta_{ij}pu_{i,j} > + < \gamma(\eta-\zeta)p >= 0,\\&< \beta q^2 > + < m_{ij}\zeta_{,i}q_{,j} > + < \gamma_{ij}qu_{i,j} > - < \gamma(\eta-\zeta)q >= 0.\end{aligned}\tag{4.98}$$

We add (4.98) equations to find

$$\begin{aligned}- < \alpha p^2 > - < \beta q^2 >=&< k_{ij}\eta_{,i}p_{,j} > + < m_{ij}\zeta_{,i}q_{,j} >\\&+ < \beta_{ij}pu_{i,j} > + < \gamma_{ij}qu_{i,j} > + < \gamma(\eta-\zeta)(p-q) > .\end{aligned}\tag{4.99}$$

Thus, we may employ (4.99) to substitute for the last five terms in (4.97) to find

$$\begin{aligned}F'' =&4 < \rho\dot{u}_i\dot{u}_i > -2 < \rho\dot{u}_i\dot{u}_i > -2 < a_{ijkh}u_{i,j}u_{k,h} >\\&-2 < \alpha p^2 > -2 < \beta q^2 >,\\=&4 < \rho\dot{u}_i\dot{u}_i > -4E(t),\\=&4 < \rho\dot{u}_i\dot{u}_i > -4E(0) + 4\int_0^t < k_{ij}p_{,i}p_{,j} > ds\\&+4\int_0^t < m_{ij}q_{,i}q_{,j} > ds + 4\int_0^t < \gamma(p-q)^2 > ds.\end{aligned}\tag{4.100}$$

Hence, since $E(0) = 0$ in this case we form $FF'' - (F')^2$ using (4.100), (4.95) and (4.96) to obtain

$$FF'' - (F')^2 = 4S^2 \geq 0,\tag{4.101}$$

where the term S^2 now has form

$$\begin{aligned} S^2 = & \Big[< \rho u_i u_i > + \int_0^t < k_{ij}\eta_{,j}\eta_{,i} > ds \\ & + \int_0^t < m_{ij}\zeta_{,j}\zeta_{,i} > ds + \int_0^t < \gamma(\eta-\zeta)^2 > ds \Big] \\ & \times \Big[< \rho \dot{u}_i \dot{u}_i > + \int_0^t < k_{ij} p_{,j} p_{,i} > ds \\ & + \int_0^t < m_{ij} q_{,j} q_{,i} > ds + \int_0^t < \gamma(p-q)^2 > ds \Big] \\ & - \Big[< \rho u_i \dot{u}_i > + \int_0^t < k_{ij}\eta_{,i} p_{,j} > ds \\ & + \int_0^t < m_{ij}\zeta_{,i} q_{,j} > ds + \int_0^t < \gamma(\eta-\zeta)(p-q) > ds \Big]^2 . \end{aligned} \tag{4.102}$$

From (4.101) we find $(\log F)'' \geq 0$ and then $F \equiv 0$, employing an argument similar to those in section 1.3.6. Thus, from the definition of F, (4.95), one sees $u_i \equiv 0$ on $\Omega \times [0,T]$. Then we appeal to (4.93) and (4.94) to find

$$0 \leq < \alpha p^2 > + < \beta q^2 > \leq 0$$

and so $p \equiv 0, q \equiv 0$ on $\Omega \times [0,T]$ and uniqueness follows.

4.3.4 Uniqueness, Double Voids

In this section we are analysing a solution to equations (4.5). To establish uniqueness we let (u_i^1, ν^1, ω^1) and (u_i^2, ν^2, ω^2) be two solutions to the boundary-initial value problem for equations (4.5) with the same source terms f_i, ℓ_1 and ℓ_2 and for the same initial and boundary data functions. The difference solution is defined as $u_i = u_i^1 - u_i^2$, $\nu = \nu^1 - \nu^2$, $\omega = \omega^1 - \omega^2$ and then one shows (u_i, ν, ω) satisfies the boundary - initial value problem

$$\begin{aligned} \rho \ddot{u}_i &= (a_{ijkh} u_{k,h})_{,j} + (b_{ij}\nu)_{,j} + (d_{ij}\omega)_{,j} , \\ \kappa_1 \ddot{\nu} &= (\alpha_{ij}\nu_{,j})_{,i} + (\beta_{ij}\omega_{,j})_{,i} - \alpha_1 \nu - \alpha_3 \omega - b_{ij} u_{i,j} , \\ \kappa_2 \ddot{\omega} &= (\gamma_{ij}\omega_{,j})_{,i} + (\beta_{ij}\nu_{,j})_{,i} - \alpha_2 \omega - \alpha_3 \nu - d_{ij} u_{i,j} , \end{aligned} \tag{4.103}$$

with the boundary conditions

$$u_i = 0, \quad \nu = 0, \quad \omega = 0, \qquad \text{on } \Gamma \times (0,T], \tag{4.104}$$

and the initial conditions

$$u_i = 0, \quad \dot{u}_i = 0, \quad \nu = 0, \quad \dot{\nu} = 0, \quad \omega = 0, \quad \dot{\omega} = 0, \qquad \mathbf{x} \in \Omega. \tag{4.105}$$

The elastic coefficients are required to satisfy the symmetry conditions (4.54) while $\alpha_{ij}, \beta_{ij}, \gamma_{ij}, b_{ij}$ and d_{ij} are symmetric.

The proof of uniqueness employs a logarithmic convexity method, cf. Iesan & Quintanilla [115]. Upon multiplying equations $(4.103)_{1,2,3}$ by $\dot{u}_i, \dot{\nu}$ and $\dot{\omega}$, respectively, then integrating over Ω using integration by parts, and then integrating in time, one may establish the energy identity

$$E(t) = E(0), \tag{4.106}$$

where $E(t)$ is defined by

$$\begin{aligned} E(t) =& \frac{1}{2} < \rho\dot{u}_i\dot{u}_i > + \frac{1}{2} < a_{ijkh}u_{k,h}u_{i,j} > + \frac{1}{2} < \kappa_1\dot{\nu}^2 > \\ &+ \frac{1}{2} < \alpha_{ij}\nu_{,i}\nu_{,j} > + \frac{1}{2} < \alpha_1\nu^2 > + \frac{1}{2} < \alpha_2\omega^2 > \\ &+ < \alpha_3\nu\omega > + \frac{1}{2} < \kappa_2\dot{\omega}^2 > + \frac{1}{2} < \gamma_{ij}\omega_{,i}\omega_{,j} > \\ &+ < b_{ij}\nu u_{i,j} > + < d_{ij}\omega u_{i,j} > + < \beta_{ij}\omega_{,j}\nu_{,i} > . \end{aligned} \tag{4.107}$$

In this case we select the function F to have form

$$F(t) = < \rho u_i u_i > + < \kappa_1\nu^2 > + < \kappa_2\omega^2 > . \tag{4.108}$$

The derivative of F is

$$F' = 2 < \rho u_i\dot{u}_i > + 2 < \kappa_1\nu\dot{\nu} > + 2 < \kappa_2\omega\dot{\omega} > . \tag{4.109}$$

The second derivative is calculated and we find

$$\begin{aligned} F'' =& 2 < \rho\dot{u}_i\dot{u}_i > + 2 < \kappa_1\dot{\nu}^2 > + 2 < \kappa_2\dot{\omega}^2 > \\ &+ 2 < \rho u_i\ddot{u}_i > + 2 < \kappa_1\nu\ddot{\nu} > + 2 < \kappa_2\omega\ddot{\omega} >, \\ =& 4\big[< \rho\dot{u}_i\dot{u}_i > + < \kappa_1\dot{\nu}^2 > + < \kappa_2\dot{\omega}^2 >\big] \\ &- 2\big[< \rho\dot{u}_i\dot{u}_i > + < \kappa_1\dot{\nu}^2 > + < \kappa_2\dot{\omega}^2 >\big] \\ &- 2 < a_{ijkh}u_{i,j}u_{k,h} > - 2 < \alpha_{ij}\nu_{,i}\nu_{,j} > - 2 < \gamma_{ij}\omega_{,i}\omega_{,j} > \\ &- 4 < b_{ij}\nu u_{i,j} > - 4 < d_{ij}\omega u_{i,j} > - 4 < \beta_{ij}\nu_{,j}\omega_{,i} > \\ &- 2 < \alpha_1\nu^2 > - 2 < \alpha_2\omega^2 > - 4 < \alpha_3\omega\nu >, \\ =& 4\big[< \rho\dot{u}_i\dot{u}_i > + < \kappa_1\dot{\nu}^2 > + < \kappa_2\dot{\omega}^2 >\big] - 4E(t), \\ =& 4\big[< \rho\dot{u}_i\dot{u}_i > + < \kappa_1\dot{\nu}^2 > + < \kappa_2\dot{\omega}^2 >\big] - 4E(0), \end{aligned} \tag{4.110}$$

where we have substituted for $\rho\ddot{u}_i$, $\kappa_1\ddot{\nu}$ and $\kappa_2\ddot{\omega}$ from (4.103) and we have also employed (4.107).

In this situation $E(0) = 0$ and so we form $FF'' - (F')^2$ to see that

$$FF'' - (F')^2 = 4S^2 \geq 0, \tag{4.111}$$

where in the present case

$$\begin{aligned} S^2 =& \big[< \rho u_i u_i > + < \kappa_1 \nu^2 > + < \kappa_2 \omega^2 >\big] \\ & \times \big[< \rho \dot{u}_i \dot{u}_i > + < \kappa_1 \dot{\nu}^2 > + < \kappa_2 \dot{\omega}^2 >\big] \\ & - \big[< \rho u_i \dot{u}_i > + < \kappa_1 \nu \dot{\nu} > + < \kappa_2 \omega \dot{\omega} >\big]^2 . \end{aligned} \tag{4.112}$$

Inequality (4.111) yields $(\log F)'' \geq 0$ and this leads to $F \equiv 0$ on $[0,T]$, cf. section 1.3.6. Then $u_i \equiv 0$, $\nu \equiv 0, \omega \equiv 0$ on $\Omega \times [0,T]$ and uniqueness follows.

In sections 4.3.1 and 4.3.2 we have seen that the single porosity and single voids models differ substantially from a mathematical viewpoint. The analysis in sections 4.3.3 and 4.3.4 displays the fact that the different terms involved in the equations for the double porosity case and for the double voids case necessitate different methods to establish uniqueness. The basic function necessary for each case is very different and the ensuing analysis proceeds differently. Thus, even though the double porosity model and the double voids model both may be used for description of an elastic body with macro and micro pores the mathematical properties of each are very different.

The fact that different analyses are required depending on whether the double porosity equations (4.87) or the double voids equations (4.103) are employed is not surprising. We see that for equations (4.87) the momentum equation involves the acceleration $\ddot{u}_i$ whereas the equations for the pressures p and q involve only first derivatives in time. The coupling terms in (4.87), i.e. those with β_{ij} and γ_{ij}, involve the pressures and the elastic velocity gradients and these terms act like a skew-symmetric operator. On the other hand, the double voids equations involve second derivatives in time of u_i, ν and ω. In fact these equations may be written in an abstract operator form as

$$A\ddot{u} = Cu \tag{4.113}$$

where A maps the Hilbert space $(L^2(\Omega))^5$ into itself and C is a densely defined operator acting on a dense subspace of $(L^2(\Omega))^5$. The operator A has the form $A = \operatorname{diag}(\rho, \rho, \rho, \kappa_1, \kappa_2)$, and acts on (u_i, ν, ω). For C we may write

$$C \begin{pmatrix} u_i \\ \nu \\ \omega \end{pmatrix} = \begin{pmatrix} (a_{ijkh} u_{k,h})_{,j} & (b_{ij} \nu)_{,j} & (d_{ij} \omega)_{,j} \\ -b_{ij} u_{i,j} & (\alpha_{ij} \nu_{,j})_{,i} - \alpha_1 \nu & (\beta_{ij} \omega_{,j})_{,i} - \alpha_3 \omega \\ -d_{ij} u_{i,j} & (\beta_{ij} \nu_{,j})_{,i} - \alpha_3 \nu & (\gamma_{ij} \omega_{,j})_{,i} - \alpha_2 \omega \end{pmatrix}$$

The elastic coefficients satisfy the symmetries (4.54) and b_{ij}, d_{ij}, α_{ij}, β_{ij} and γ_{ij} are symmetric tensors. To verify that C is a symmetric linear operator we let $u = (u_i, \nu, \omega)^T$ and $v = (v_i, \phi, \psi)$, and then we form the inner products in $(L^2(\Omega))^5$ of (v, Cu) and (u, Cv), with u_i, ν, ω, v_i, ϕ, ψ equal to zero on Γ. After integration by parts and use of the boundary conditions one may verify that

$$
\begin{aligned}
(v,Cu) = &< v_i(a_{ijkh}u_{k,h})_{,j} > + < v_i(b_{ij}\nu)_{,j} > + < v_i(d_{ij}\omega)_{,j} > \\
&+ < \phi(\alpha_{ij}\nu_{,j})_{,i} > + < \phi(\beta_{ij}\omega_{,j})_{,i} > - < \alpha_1\nu\phi > \\
&- < \alpha_3\omega\phi > - < b_{ij}u_{i,j}\phi > \\
&+ < \psi(\gamma_{ij}\omega_{,j})_{,i} > + < \psi(\beta_{ij}\nu_{,j})_{,i} > - \alpha_2\omega\psi > \\
&- < \alpha_3\nu\psi > - < d_{ij}u_{i,j}\psi > \\
=&- < a_{ijkh}u_{i,j}v_{k,h} > - < \nu b_{ij}v_{i,j} > - < \omega d_{ij}v_{i,j} > \\
&- < \alpha_{ij}\phi_{,j}\nu_{,i} > - < \beta_{ij}\omega_{,i}\phi_{,j} > - < \alpha_1\phi\nu > - < \alpha_3\phi\omega > \\
&+ < u_i(b_{ij}\phi)_{,j} > - < \gamma_{ij}\omega_{,i}\psi_{,j} > - < \beta_{ij}\psi_{,j}\nu_{,i} > - < \alpha_2\psi\omega > \\
&- < \alpha_3\psi\nu > + < u_i(d_{ij}\psi)_{,j} > \\
= &< u_i(a_{ijkh}v_{k,h})_{,j} > + < u_i(b_{ij}\phi)_{,j} > + < u_i(d_{ij}\psi)_{,j} > \\
&+ < \nu(\alpha_{ij}\phi_{,j})_{,i} > + < \nu(\beta_{ij}\psi_{,j})_{,i} > - < \alpha_1\phi\nu > - < \alpha_3\phi\omega > \\
&- < b_{ij}v_{i,j}\nu > + < \omega(\gamma_{ij}\psi_{,j})_{,i} > + < \omega(\beta_{ij}\phi_{,j})_{,i} > \\
&< \alpha_2\psi\omega > - < \alpha_3\psi\nu > - < \alpha_{ij}v_{i,j}\omega > \\
=&(u,Cv).
\end{aligned}
$$

Thus, the operator C in equation (4.113) applied to the double voids theory (4.103) is a symmetric one.

4.4 Exercises

Exercise 4.1. Use the Cauchy-Schwarz inequality to show $S^2 \geq 0$, where S^2 is defined in (4.71).

Exercise 4.2. Construct a rigorous uniqueness proof along the lines of that given in section 1.3.6 by adding ε to $F(t)$ in (4.59).

Exercise 4.3. Use the Cauchy-Schwarz inequality to show $S^2 \geq 0$, where S^2 is defined in (4.86).

Exercise 4.4. Construct a rigorous uniqueness proof along the lines of that given in section 1.3.6 by adding ε to $F(t)$ in (4.78).

Exercise 4.5. Use the Cauchy-Schwarz inequality to show $S^2 \geq 0$, where S^2 is defined in (4.102).

Exercise 4.6. Construct a rigorous uniqueness proof along the lines of that given in section 1.3.6 by adding ε to $F(t)$ in (4.95).

Exercise 4.7. Use the Cauchy-Schwarz inequality to show $S^2 \geq 0$, where S^2 is defined in (4.112).

Exercise 4.8. Construct a rigorous uniqueness proof along the lines of that given in section 1.3.6 by adding ε to $F(t)$ in (4.108).

Exercise 4.9. Let (u_i^1, v^1) be a solution to equations (4.3) with the boundary conditions (4.73) and the initial conditions (4.74). Let (u_i^2, v^2) be another solution to equations (4.3) with the boundary conditions (4.73) for the same data functions, although this solution satisfies the initial conditions (4.74) for a different set of data functions. Suppose $\rho(\mathbf{x}) > 0$, $\chi_1(\mathbf{x}) > 0$, $b_{ij}(\mathbf{x}), \alpha_{ij}(\mathbf{x})$ are symmetric, $\xi(\mathbf{x}) > 0$, and suppose the function

$$\begin{aligned} V(t) =& \frac{1}{2} < a_{ijkh} u_{i,j} u_{k,h} > + \frac{1}{2} < \alpha_{ij} v_{,i} v_{,j} > \\ &+ \frac{1}{2} < \xi v^2 > + < b_{ij} v u_{i,j} > \end{aligned}$$

is such that

$$V(t) \geq k_1 \|\nabla \mathbf{u}\|^2 + k_2 \|\nabla v\|^2 \tag{4.114}$$

$k_1, k_2 > 0$ constants, for any solution to (4.3) which has $u_i \equiv 0$, $v \equiv 0$ on Γ. Suppose also a_{ijkh} satisfy the symmetries

$$a_{ijkh} = a_{jikh} = a_{khij} \, . \tag{4.115}$$

Show that (u_i, v) depends continuously upon the initial data for all time.
Hint. Let $u_i = u_i^1 - u_i^2, v = v^1 - v^2$. Show that the energy $E(t)$ given by

$$\begin{aligned} E(t) =& \frac{1}{2} < \rho \dot{u}_i \dot{u}_i > + \frac{1}{2} < a_{ijkh} u_{i,j} u_{k,h} > + \frac{1}{2} < \chi_1 \dot{v}^2 > \\ &+ \frac{1}{2} < \alpha_{ij} v_{,i} v_{,j} > + \frac{1}{2} < \xi v^2 > + < b_{ij} v u_{i,j} > \end{aligned}$$

satsifies the equation

$$E(t) = E(0).$$

The use (4.114) to establish

$$\frac{1}{2} < \rho \dot{u}_i \dot{u}_i > + \frac{1}{2} < \chi_1 \dot{v}^2 > + k_1 \|\nabla \mathbf{u}\|^2 + k_2 \|\nabla v\|^2 \leq E(0).$$

Continuous dependence on the initial data follows from this inequality with the aid of Poincaré's inequality.

Exercise 4.10. Let (u_i^1, v^1), (u_i^2, v^2) be as in exercise 4.9 but do not require (4.114). Instead require a_{ijkh} to only satisfy (4.115). Show that if u_i and v are bounded in $L^2(\Omega)$ norm for $t \in [0,T]$, some $T < \infty$, then the solution depends continuously on the initial data in the sense of Hölder stability on compact sub-intervals of $[0,T)$.
Hint. Use a logarithmic convexity argument following the outline for classical elastodynamics in sections 1.4.4 - 1.4.6. Use for the function F

$$F(t) = < \rho u_i u_i > + < \chi_1 v^2 >$$

when $E(0) \leq 0$, where $E(t)$ is given in exercise 4.9. When $E(0) > 0$ show that the function

$$G(t) = \log\left[F(t) + 2E(0)\right] + t^2$$

is convex and follow similar arguments to those in section 1.4.6.

Exercise 4.11. Let (u_i, v) be a solution to equations (4.3) with the boundary conditions as in exercise 4.9. Assume the symmetries (4.115) but impose no definiteness on a_{ijkh}. Let the initial conditions $u_i(\mathbf{x},0), \dot{u}_i(\mathbf{x},0), v(\mathbf{x},0)$ and $\dot{v}(\mathbf{x},0)$ be non-zero. Define $F(t)$ as in the hint for exercise 4.10. Show that $F(t)$ will have exponential growth if $E(0) \leq 0$, or if $E(0) > 0$ and $F'(0) > 0$.
Hint. The calculations from (4.78) to (4.84) should help, along with exercise 1.14 in chapter 1.

Chapter 5
Uniqueness and Stability by Energy Methods

In this chapter we investigate questions of uniqueness, continuous dependence on the initial data, and continuous dependence on the model itself, by employing techniques based on the energy function generated by the system of equations. We concentrate on equations of elasticity with a double porosity structure, or thermoelasticity with a double porosity structure. We firstly deal with a thermoelastic body with a double porosity structure allowing for cross inertia coefficients, equations (2.33), when the body Ω is bounded. After that we examine the equations for isothermal elasticity with double porosity, equations (2.25), (2.26) and (2.27), when the domain Ω is unbounded.

5.1 Uniqueness in double porosity thermoelasticity

We begin by establishing sufficient conditions to demonstrate uniqueness for a solution to equations (2.33) which govern the behaviour of a thermoelastic body with a double porosity structure. As in section 2.3.1 let $u_i(\mathbf{x},t)$ denote the elastic displacement, $p(\mathbf{x},t)$ denote the pressure in the macro pores, $q(\mathbf{x},t)$ denote the pressure in the micro pores, and let $\theta(\mathbf{x},t)$ denote the temperature in the body. Then, the linear anisotropic equations for a double porosity thermoelastic material may be written as, cf. equations (2.33),

$$\begin{aligned}
\rho\ddot{u}_i &= (a_{ijkh}u_{k,h})_{,j} - (\beta_{ij}p)_{,j} - (\gamma_{ij}q)_{,j} - (a_{ij}\theta)_{,j} + \rho f_i\,,\\
\alpha\dot{p} + \alpha_1\dot{q} + \alpha_2\dot{\theta} &= (k_{ij}p_{,j})_{,i} - \beta_{ij}\dot{u}_{i,j} - \gamma(p-q) + \rho s_1\,,\\
\alpha_1\dot{p} + \beta\dot{q} + \beta_1\dot{\theta} &= (m_{ij}q_{,j})_{,i} - \gamma_{ij}\dot{u}_{i,j} + \gamma(p-q) + \rho s_2\,,\\
\alpha_2\dot{p} + \beta_1\dot{q} + a\dot{\theta} &= (r_{ij}\theta_{,j})_{,i} - a_{ij}\dot{u}_{i,j} + \rho r\,.
\end{aligned} \tag{5.1}$$

In these equations $\rho = \rho(\mathbf{x}) > 0$ is the density, $a_{ijkh}(\mathbf{x})$ are the elastic coefficients, $k_{ij}, m_{ij}, r_{ij}, \beta_{ij}, \gamma_{ij}$ and a_{ij} are all symmetric tensors. The inertia coefficients $\alpha, \alpha_1, \alpha_2, \beta, \beta_1$ and a and the interaction coefficient γ all depend on $\mathbf{x}$. The elastic

B. Straughan, *Mathematical Aspects of Multi–Porosity Continua*, Advances in Mechanics and Mathematics 38, https://doi.org/10.1007/978-3-319-70172-1_5

coefficients are required to be symmetric in the sense of (1.16), namely

$$a_{ijkh} = a_{khij} = a_{jikh}. \tag{5.2}$$

The term $f_i(\mathbf{x},t)$ represents a prescribed body force, $s_1(\mathbf{x},t), s_2(\mathbf{x},t)$ are source terms for the macro and micro pressure equations, and $r(\mathbf{x},t)$ is a prescribed heat source. The effect of a heat source may be important in double porosity thermoelasticity as it could make a strong contribution to inducing internal thermal stresses.

As elsewhere in this book we let Ω denote a bounded domain in $\mathbb{R}^3$ with boundary Γ sufficiently smooth to allow application of the divergence theorem. We shall now derive sufficient conditions to establish uniqueness of a solution to the fundamental equations (5.1). We suppose that (u_i, p, q, θ) is a solution to equations (5.1) in the domain $\Omega \times (0,T)$ where $T < \infty$ is a fixed time. This solution is required to satisfy the boundary conditions

$$\begin{aligned} &u_i(\mathbf{x},t) = u_i^B(\mathbf{x},t), \qquad p(\mathbf{x},t) = p^B(\mathbf{x},t), \qquad q(\mathbf{x},t) = q^B(\mathbf{x},t), \\ &\theta(\mathbf{x},t) = \theta^B(\mathbf{x},t), \qquad \mathbf{x} \in \Gamma \times (0,T), \end{aligned} \tag{5.3}$$

where u_i^B, p^B, q^B, s^B, and θ^B are prescribed functions. In addition, we require the solution (u_i, p, q, θ) to satisfy the initial conditions

$$\begin{aligned} &u_i(\mathbf{x},0) = v_i(\mathbf{x}), \qquad \dot{u}_i(\mathbf{x},0) = w_i(\mathbf{x}), \qquad p(\mathbf{x},0) = P(\mathbf{x}), \\ &q(\mathbf{x},0) = Q(\mathbf{x}), \qquad \theta(\mathbf{x},0) = \Theta(\mathbf{x}), \qquad \mathbf{x} \in \Omega, \end{aligned} \tag{5.4}$$

where the functions v_i, w_i, P, Q and Θ are prescribed. Let us denote the boundary - initial value problem for equations (5.1) together with (5.3) and (5.4) by $\mathscr{P}$.

in this section we suppose that the coefficients k_{ij}, m_{ij} and r_{ij} are functions of $\mathbf{x}$ which additionally are non-negative, i.e.

$$k_{ij}\xi_i\xi_j \geq 0, \qquad m_{ij}\xi_i\xi_j \geq 0, \qquad r_{ij}\xi_i\xi_j \geq 0, \tag{5.5}$$

for all ξ_i. The coupling coefficients β_{ij}, γ_{ij} and a_{ij} are functions of $\mathbf{x}$ while the interaction coefficient γ is a positive function of $\mathbf{x}$. The inertia coefficients $\alpha, \beta, a, \alpha_1, \alpha_3$ and β_2 are functions of the spatial variable $\mathbf{x}$. The inertia coefficients are such that $\alpha > 0, \beta > 0$ and $a > 0$. We also require that the symmetric matrix

$$A = \begin{pmatrix} \alpha & \alpha_1 & \alpha_2 \\ \alpha_1 & \beta & \beta_1 \\ \alpha_2 & \beta_1 & a \end{pmatrix}$$

is positive - definite. This means that there exist positive constants k_1, k_2, k_3 such that

$$\mathbf{y}^T A \mathbf{y} \geq k_1^2 y_1^2 + k_2 y_2^2 + k_3 y_3^2, \tag{5.6}$$

for a typical vector $\mathbf{y} = (p, q, \theta)$, i.e.

$$(p,q,\theta)\begin{pmatrix}\alpha & \alpha_1 & \alpha_2\\ \alpha_1 & \beta & \beta_1\\ \alpha_2 & \beta_1 & a\end{pmatrix}\begin{pmatrix}p\\ q\\ \theta\end{pmatrix} \geq k_1^2p^2+k_2^2q^2+k_3^2\theta^2 .$$

In sections 5.1, 5.2, 5.2.1 and 5.2.2 we shall suppose that the elastic coefficients a_{ijkh} are functions of $\mathbf{x}$ and

$$a_{ijkh}\xi_{ij}\xi_{kh} \geq a_0\xi_{ij}\xi_{ij} \tag{5.7}$$

for all ξ_{ij}. When we deal with the uniqueness question we set $a_0 = 0$, but for continuous dependence (stability) we suppose that $a_0 > 0$.

To establish uniqueness for a solution to $\mathscr{P}$ we let the functions (u_i^1,p^1,q^1,θ^1) and (u_i^2,p^2,q^2,θ^2) be two solutions to $\mathscr{P}$ for the same boundary data functions h_i^B,p^B,q^B,θ^B, for the same initial data functions v_i,w_i,P,Q,Θ, and for the same source terms f_i,s^1,s^2,r. Define the difference variables u_i,p,q and θ by the relations

$$u_i = u_i^1-u_i^2, \qquad p=p^1-p^1, \qquad q=q^1-q^2, \qquad \theta=\theta^1-\theta^2. \tag{5.8}$$

Then by subtraction, one may verify from equations (5.1), (5.3) and (5.4), that the difference solution (u_i,p,q,θ) satisfies the boundary-initial value problem

$$\begin{aligned}
&\rho\ddot{u}_i = (a_{ijkh}u_{k,h})_{,j}-(\beta_{ij}p)_{,j}-(\gamma_{ij}q)_{,j}-(a_{ij}\theta)_{,j}\,,\\
&\alpha\dot{p}+\alpha_1\dot{q}+\alpha_2\dot{\theta} = (k_{ij}p_{,j})_{,i}-\gamma(p-q)-\beta_{ij}\dot{u}_{i,j}\,,\\
&\alpha_1\dot{p}+\beta\dot{q}+\beta_1\dot{\theta} = (m_{ij}q_{,j})_{,i}+\gamma(p-q)-\gamma_{ij}\dot{u}_{i,j}\,,\\
&\alpha_2\dot{p}+\beta_1\dot{q}+a\dot{\theta} = (r_{ij}\theta_{,j})_{,i}-a_{ij}\dot{u}_{i,j}\,,
\end{aligned} \tag{5.9}$$

with the boundary conditions

$$u_i(\mathbf{x},t)=0, \qquad p(\mathbf{x},t)=0, \qquad q(\mathbf{x},t)=0, \qquad \theta(\mathbf{x},t)=0, \qquad \mathbf{x}\in\Gamma, \tag{5.10}$$

and the initial conditions

$$\begin{aligned}
&u_i(\mathbf{x},0)=0, \qquad \dot{u}_i(\mathbf{x},0)=0, \qquad p(\mathbf{x},0)=0,\\
&q(\mathbf{x},0)=0, \qquad \theta(\mathbf{x},0)=0, \qquad \mathbf{x}\in\Omega.
\end{aligned} \tag{5.11}$$

To establish uniqueness of a solution to $\mathscr{P}$ with conditions (5.5), (5.6) and (5.7) we multiply equation $(5.9)_1$ by $\dot{u}_i$ and integrate over Ω. Then multiply equations $(5.9)_2$ - $(5.9)_4$, respectively, by p,q and θ and integrate each over Ω. Integrate by parts, use the boundary conditions, and add the resulting equations together. Define the energy function $E(t)$ by

$$\begin{aligned}
E(t) =&\frac{1}{2}<\rho\dot{u}_i\dot{u}_i>+\frac{1}{2}<a_{ijkh}u_{i,j}u_{k,h}>+\frac{1}{2}<\alpha p^2>+\frac{1}{2}<\beta q^2>\\
&+\frac{1}{2}<a\theta^2>+<\alpha_1 pq>+<\alpha_2\theta p>+<\beta_1 q\theta>.
\end{aligned} \tag{5.12}$$

The result of adding the equations formed as outlined above is

$$
\begin{aligned}
E(t) &+ \int_0^t < k_{ij}p_{,i}p_{,j} > da + \int_0^t < m_{ij}q_{,i}q_{,j} > da \\
&+ \int_0^t < r_{ij}\theta_{,i}\theta_{,j} > da + \int_0^t < \gamma(p-q)^2 > da = E(0),
\end{aligned} \tag{5.13}
$$

where da denotes integration with respect to time.

For uniqueness $E(0) = 0$. Thus, from (5.13), one may use conditions (5.5), (5.6) and (5.7) to see that

$$
0 \leq < \rho \dot{u}_i \dot{u}_i > + k_1^2\|p\|^2 + k_2^2\|q\|^2 + k_3^2\|\theta\|^2 \leq 0.
$$

Hence

$$
p \equiv 0, \qquad q \equiv 0, \qquad \theta \equiv 0, \qquad \text{and} \quad \dot{u}_i \equiv 0, \quad \text{in } \Omega \times (0,T).
$$

Since $\dot{u}_i \equiv 0$ and $u_i = 0$ at $t = 0$ it follows that $u_i \equiv 0$ in $\Omega \times (0,T)$. Thus, uniqueness of a solution to $\mathscr{P}$ follows.

5.2 Continuous dependence upon the initial data

The first analysis now given is for stability in the sense of continuous dependence upon the initial data. Thus, suppose there are two solutions $(u_i^1, p^1, q^1, \theta^1)$ and $(u_i^2, p^2, q^2, \theta^2)$ which satisfy equations (5.1) for the same body force f_i, for the same source functions s^1 and s^2, and for the same heat source r. These solutions also satisfy the boundary conditions (5.3) for the same functions u_i^B, p^B, q^B and θ^B. However, the two solutions are subject to different initial conditions in that $(u_i^1, p^1, q^1, \theta^1)$ is required to satisfy the initial conditions

$$
\begin{aligned}
&u_i^1(\mathbf{x},0) = v_i^1(\mathbf{x}), \qquad \dot{u}_i^1(\mathbf{x},0) = w_i^1(\mathbf{x}), \qquad p^1(\mathbf{x},0) = P^1(\mathbf{x}), \\
&q^1(\mathbf{x},0) = Q^1(\mathbf{x}), \qquad \theta^1(\mathbf{x},0) = \Theta^1(\mathbf{x}),
\end{aligned} \tag{5.14}
$$

for $\mathbf{x} \in \Omega$, whereas $(u_i^2, p^2, q^2, \theta^2)$ is required to satisfy

$$
\begin{aligned}
&u_i^2(\mathbf{x},0) = v_i^2(\mathbf{x}), \qquad \dot{u}_i^2(\mathbf{x},0) = w_i^2(\mathbf{x}), \qquad p^2(\mathbf{x},0) = P^2(\mathbf{x}), \\
&q^2(\mathbf{x},0) = Q^2(\mathbf{x}), \qquad \theta^2(\mathbf{x},0) = \Theta^2(\mathbf{x}),
\end{aligned} \tag{5.15}
$$

for $\mathbf{x} \in \Omega$. In equations (5.14) and (5.15) the functions $v_i^1(\mathbf{x}), \ldots, \Theta^2(\mathbf{x})$ are given.

Now form the difference solution (u_i, p, q, θ) as in (5.8). The difference solution satisfies the differential equations (5.9) together with the boundary conditions (5.10). Define the data functions v_i, w_i, P, Q and Θ by

$$\begin{aligned} &v_i(\mathbf{x}) = v_i^1(\mathbf{x}) - v_i^2(\mathbf{x}), \quad w_i(\mathbf{x}) = w_i^1(\mathbf{x}) - w_i^2(\mathbf{x}), \\ &P(\mathbf{x}) = P^1(\mathbf{x}) - P^2(\mathbf{x}), \quad Q(\mathbf{x}) = Q^1(\mathbf{x}) - Q^2(\mathbf{x}), \\ &\Theta(\mathbf{x}) = \Theta^1(\mathbf{x}) - \Theta^2(\mathbf{x}), \qquad \mathbf{x} \in \Omega. \end{aligned} \tag{5.16}$$

The difference solution (u_i, p, q, θ) satisfies the initial conditions

$$\begin{aligned} &u_i(\mathbf{x},0) = v_i(\mathbf{x}), \qquad \dot{u}_i(\mathbf{x},0) = w_i(\mathbf{x}), \qquad p(\mathbf{x},0) = P(\mathbf{x}), \\ &q(\mathbf{x},0) = Q(\mathbf{x}), \qquad \theta(\mathbf{x},0) = \Theta(\mathbf{x}), \qquad \mathbf{x} \in \Omega. \end{aligned} \tag{5.17}$$

Thus, the difference solution (u_i, p, q, θ) satisfies the boundary initial value problem $\mathscr{P}_1$ composed of equations (5.9), (5.10) and (5.17).

By an analogous procedure to that leading to equation (5.13) one finds that this equation still holds. Now, however, $E(0) > 0$. Using inequality (5.5) we, therefore, deduce

$$E(t) \le E(0), \tag{5.18}$$

where

$$\begin{aligned} E(0) =& \frac{1}{2} < \rho w_i w_i > + \frac{1}{2} < a_{ijkh} v_{i,j} v_{k,h} > + \frac{1}{2} < \alpha P^2 > + \frac{1}{2} < \beta Q^2 > \\ &+ \frac{1}{2} < a\Theta^2 > + < \alpha_1 PQ > + < \alpha_2 \Theta P > + < \beta_1 Q \Theta > . \end{aligned} \tag{5.19}$$

Now use inequality (5.6) in inequality (5.18) to derive

$$\frac{1}{2} < \rho \dot{u}_i \dot{u}_i > + \frac{1}{2} < a_{ijkh} u_{i,j} u_{k,h} > + k_1 \|p\|^2 + k_2 \|q\|^2 + k_3 \|\theta\|^2 \le E(0). \tag{5.20}$$

Finally, employ inequality (5.7) for a constant $a_0 > 0$ to obtain from inequality (5.20)

$$\frac{1}{2} < \rho \dot{u}_i \dot{u}_i > + \frac{a_0}{2} \|\nabla \mathbf{u}\|^2 + k_1 \|p\|^2 + k_2 \|q\|^2 + k_3 \|\theta\|^2 \le E(0). \tag{5.21}$$

Inequality (5.21) establishes continuous dependence on the initial data for all $t > 0$. This continuous dependence is in the L^2 measure for $\dot{\mathbf{u}}, p, q$ and θ and in the $H_0^1(\Omega)$ measure for $\mathbf{u}$.

5.2.1 Continuous dependence upon the body force

We now consider continuous dependence of the solution to equations (5.1) upon changes in the body force, f_i. This introduces another class of stability problem, namely where one considers a change to the model itself. This introduces an important area known as structural stability. Many writers believe structural stability is as important a concept as continuous dependence on the initial data. It is noteworthy that structural stability features prominently in the classic text of Hirsch & Smale [104]. Structural stability was analysed in depth by Knops & Payne [141] for a variety of topics in the field of linear elastodynamics. Improvements to the work of Knops & Payne [141] are included in the work of Knops & Payne [144]. Structural stability is analysed in detail in chapter 2 of the book by Straughan [202].

To investigate continuous dependence on the body force for a solution to equations (5.1) we let $(u_i^1, p^1, q^1, \theta^1)$ be a solution to equations (5.1) for a body force $f_i^1(\mathbf{x},t)$ and we let $(u_i^2, p^2, q^2, \theta^2)$ be a solution to equations (5.1) for a body force $f_i^2(\mathbf{x},t)$. We suppose that both solutions are subject to the same source functions $s^1(\mathbf{x},t), s^2(\mathbf{x},t), r(\mathbf{x},t)$ and both solutions satisfy the boundary conditions (5.3) and the initial conditions (5.4) for the *same* data functions $u_i^B, p^B, q^B, \theta^B$ and v_i, w_i, P, Q and Θ.

Define the difference body force f_i by

$$f_i(\mathbf{x},t) = f_i^1(\mathbf{x},t) - f_i^2(\mathbf{x},t). \tag{5.22}$$

One may verify that the difference solution defined by (5.8) satisfies the boundary-initial value problem

$$\begin{aligned}
\rho \ddot{u}_i &= (a_{ijkh}u_{k,h})_{,j} - (\beta_{ij}p)_{,j} - (\gamma_{ij}q)_{,j} - (a_{ij}\theta)_{,j} + \rho f_i\,,\\
\alpha \dot{p} + \alpha_1 \dot{q} + \alpha_2 \dot{\theta} &= (k_{ij}p_{,j})_{,i} - \gamma(p-q) - \beta_{ij}\dot{u}_{i,j}\,,\\
\alpha_1 \dot{p} + \beta \dot{q} + \beta_1 \dot{\theta} &= (m_{ij}q_{,j})_{,i} + \gamma(p-q) - \gamma_{ij}\dot{u}_{i,j}\,,\\
\alpha_2 \dot{p} + \beta_1 \dot{q} + a\dot{\theta} &= (r_{ij}\theta_{,j})_{,i} - a_{ij}\dot{u}_{i,j}\,,
\end{aligned} \tag{5.23}$$

with the boundary conditions

$$u_i(\mathbf{x},t) = 0, \qquad p(\mathbf{x},t) = 0, \qquad q(\mathbf{x},t) = 0, \qquad \theta(\mathbf{x},t) = 0, \qquad \mathbf{x} \in \Gamma, \tag{5.24}$$

and the initial conditions

$$\begin{aligned}
u_i(\mathbf{x},0) &= 0, \qquad \dot{u}_i(\mathbf{x},0) = 0, \qquad p(\mathbf{x},0) = 0,\\
q(\mathbf{x},0) &= 0, \qquad \theta(\mathbf{x},0) = 0, \qquad \mathbf{x} \in \Omega.
\end{aligned} \tag{5.25}$$

To investigate continuous dependence upon the body force we again use an energy method. To begin we multiply equation $(5.23)_1$ by $\dot{u}_i$ and integrate over Ω. We then multiply equations $(5.23)_2$ - $(5.23)_4$ each, respectively, by p, q and θ and integrate over Ω. After using the boundary conditions (5.24) and adding the results one may show find that

$$\begin{aligned}
&\frac{dE}{dt} + < k_{ij}p_{,i}p_{,j} > + < m_{ij}q_{,i}q_{,j} >\\
&\quad + < r_{ij}\theta_{,i}\theta_{,j} > + < \gamma(p-q)^2 > = < \rho f_i \dot{u}_i >,
\end{aligned} \tag{5.26}$$

where $E(t)$ is the energy function which is the same as that defined in (5.12). Now use the arithmetic - geometric mean inequality on the term on the right of (5.26) to see that

$$<\rho f_i\dot{u}_i> \leq \frac{1}{2} <\rho\dot{u}_i\dot{u}_i> + \frac{1}{2} <\rho f_i f_i> . \tag{5.27}$$

Next, employ conditions (5.5) drop some terms on the left hand side of (5.26) and then after using (5.27) in (5.26) one uses the integrating factor e^{-t} and integrates in t to derive

$$\begin{aligned}\frac{1}{2} <\rho\dot{u}_i\dot{u}_i> &+ \frac{1}{2} <a_{ijkh}u_{i,j}u_{k,h}> + \frac{1}{2}k_1\|p\|^2 + \frac{1}{2}k_2\|q\|^2 \\ &+ \frac{1}{2}k_3\|\theta\|^2 \leq \int_0^t e^{(t-a)} <\rho f_i(\mathbf{x},a)f_i(\mathbf{x},a)> da.\end{aligned} \tag{5.28}$$

Use of inequality (5.7) with $a_0 > 0$ allows one to obtain the inequality

$$\begin{aligned}\frac{1}{2} <\rho\dot{u}_i\dot{u}_i> &+ \frac{1}{2}a_0\|\nabla\mathbf{u}\|^2 + \frac{1}{2}k_1\|p\|^2 + \frac{1}{2}k_2\|q\|^2 \\ &+ \frac{1}{2}k_3\|\theta\|^2 < \int_0^t e^{(t-a)} <\rho f_i(\mathbf{x},a)f_i(\mathbf{x},a)> da.\end{aligned} \tag{5.29}$$

Inequality (5.29) establishes continuous dependence of a solution to equations (5.1) upon the body force f_i, in the measure indicated on the left hand side of (5.29).

5.2.2 Continuous dependence upon the heat source

We now investigate another problem of structural stability, namely, continuous dependence of a solution to equations (5.1) upon changes in the heat source $r(\mathbf{x},t)$.

The analysis progresses as in section 5.2.1 and we define $(u_i^1, p^1, q^1, \theta^1)$ and $(u_i^2, p^2, q^2, \theta^2)$ to be two solutions to equations (5.1); however, they now have the *same* sources f_i, s^1, s^2, but they have different heat sources r^1 and r^2, respectively. Both solutions satisfy the same boundary and initial conditions. The difference solution defined by (5.8) satisfies the boundary conditions (5.24) and initial conditions (5.25). The difference solution (u_i, p, q, θ) in this case satisfies the partial differential equations

$$\begin{aligned}\rho\ddot{u}_i &= (a_{ijkh}u_{k,h})_{,j} - (\beta_{ij}p)_{,j} - (\gamma_{ij}q)_{,j} - (a_{ij}\theta)_{,j} ,\\ \alpha\dot{p} + \alpha_1\dot{q} + \alpha_2\dot{\theta} &= (k_{ij}p_{,j})_{,i} - \gamma(p-q) - \beta_{ij}\dot{u}_{i,j} ,\\ \alpha_1\dot{p} + \beta\dot{q} + \beta_1\dot{\theta} &= (m_{ij}q_{,j})_{,i} + \gamma(p-q) - \gamma_{ij}\dot{u}_{i,j} ,\\ \alpha_2\dot{p} + \beta_1\dot{q} + a\dot{\theta} &= (r_{ij}\theta_{,j})_{,i} - a_{ij}\dot{u}_{i,j} + \rho r .\end{aligned} \tag{5.30}$$

We again use an energy technique and multiply each of $(5.30)_1$ - $(5.30)_4$ by $\dot{u}_i, p, q$ and θ, respectively, and integrate over Ω. This now leads to an energy equation of form

$$\begin{aligned}\frac{dE}{dt} &+ <k_{ij}p_{,i}p_{,j}> + <m_{ij}q_{,i}q_{,j}> \\ &+ <r_{ij}\theta_{,i}\theta_{,j}> + <\gamma(p-q)^2> = <\rho\theta r> .\end{aligned} \tag{5.31}$$

In this section we require k_{ij} and m_{ij} to satisfy (5.5) but now we suppose that r_{ij} satisfy the condition

$$r_{ij}\xi_i\xi_j \geq r_0\xi_i\xi_i, \tag{5.32}$$

for all ξ_i, and for a constant $r_0 > 0$. We employ inequality (5.32) in (5.31) and then we use the arithmetic - geometric mean inequality to obtain

$$\frac{dE}{dt} + r_0\|\nabla\theta\|^2 \leq \frac{1}{2b} < \theta^2 > + \frac{b}{2} < \rho^2 r^2 >, \tag{5.33}$$

for a constant $b > 0$ at our disposal. We next employ Poincaré's inequality in the form $\|\nabla\theta\|^2 \geq \lambda_1\|\theta\|^2$, cf. (1.10). Select the constant b as $b = 1/2r_0\lambda_1$ and then from (5.33) one finds

$$\frac{dE}{dt} \leq \frac{b}{2} < \rho^2 r^2 > . \tag{5.34}$$

This inequality is integrated in time and we use inequality (5.6) to see that

$$\begin{aligned} < \rho\dot{u}_i\dot{u}_i > + < a_{ijkh}u_{i,j}u_{k,h} > &+ k_1\|p\|^2 + k_2\|q\|^2 \\ &+ k_3\|\theta\|^2 \leq b\int_0^t < \rho^2 r^2 > da. \end{aligned} \tag{5.35}$$

Now use inequality (5.7) with $a_0 > 0$ to show that

$$\begin{aligned} < \rho\dot{u}_i\dot{u}_i > + a_0\|\nabla\mathbf{u}\|^2 &+ k_1\|p\|^2 + k_2\|q\|^2 \\ &+ k_3\|\theta\|^2 \leq \frac{1}{2r_0\lambda_1}\int_0^t < \rho^2 r^2 > da. \end{aligned} \tag{5.36}$$

Inequality (5.36) is an *a priori* estimate which demonstrates continuous dependence of a solution to the double porosity thermoelastic equations (5.1) upon changes in the heat source $r(\mathbf{x},t)$.

One can also establish continuous dependence upon the source terms s_1 and s_2, see exercise 5.2.

5.3 Uniqueness on an unbounded domain

In this section we let Ω be a domain exterior to a bounded domain $\Omega_0 \subset \mathbb{R}^3$, as in section 1.3.5. The interior boundary of Ω is Γ. Our aim is to prove a uniqueness theorem for a solution to the isothermal double porosity equations (2.25), (2.26) and (2.27) without requiring decay of the solution as $|\mathbf{x}| = r \to \infty$.

Let, therefore, $\mathscr{P}$ denote the boundary-initial value problem

$$\begin{aligned} \rho\ddot{u}_i &= (a_{ijkh}u_{k,h})_{,j} - (\beta_{ij}p)_{,j} - (\gamma_{ij}q)_{,j} + \rho f_i, \\ \alpha\dot{p} &= (k_{ij}p_{,j})_{,i} - \gamma(p-q) - \beta_{ij}\dot{u}_{i,j} + \rho s^1, \\ \beta\dot{q} &= (m_{ij}q_{,j})_{,i} + \gamma(p-q) - \gamma_{ij}\dot{u}_{i,j} + \rho s^2, \end{aligned} \tag{5.37}$$

in $\Omega \times \{t > 0\}$, together with the boundary conditions

$$u_i = u_i^B(\mathbf{x},t), \quad p = p^B(\mathbf{x},t), \quad q = q^B(\mathbf{x},t), \qquad \text{on } \Gamma \times \{t > 0\}, \tag{5.38}$$

and the initial conditions

$$\begin{aligned} &u_i(\mathbf{x},0) = v_i(\mathbf{x}), \quad \dot{u}_i(\mathbf{x},0) = w_i(\mathbf{x}), \\ &p(\mathbf{x},0) = P(\mathbf{x}), \quad q(\mathbf{x},0) = Q(\mathbf{x}), \qquad \mathbf{x} \in \Omega. \end{aligned} \tag{5.39}$$

To analyse the uniqueness question we let (u_i^1, p^1, q^1) and (u_i^2, p^2, q^2) be solutions to $\mathscr{P}$ for the same boundary and initial data $u_i^B, p^B, q^B, v_i, w_i, P$ and Q and for the same source functions f_i, s^1, s^2. Define the difference variables u_i, p and q by

$$u_i = u_i^1 - u_i^2, \qquad p = p^1 - p^2, \qquad q = q^1 - q^2$$

and then the difference solution satisfies the boundary - initial value problem comprised of

$$\begin{aligned} &\rho \ddot{u}_i = (a_{ijkh} u_{k,h})_{,j} - (\beta_{ij} p)_{,j} - (\gamma_{ij} q)_{,j}, \\ &\alpha \dot{p} = (k_{ij} p_{,j})_{,i} - \gamma(p - q) - \beta_{ij} \dot{u}_{i,j}, \\ &\beta \dot{q} = (m_{ij} q_{,j})_{,i} + \gamma(p - q) - \gamma_{ij} \dot{u}_{i,j}, \end{aligned} \tag{5.40}$$

in $\Omega \times \{t > 0\}$, together with the boundary conditions

$$u_i = 0, \quad p = 0, \quad q = 0, \qquad \text{on } \Gamma \times \{t > 0\}, \tag{5.41}$$

and the initial conditions

$$u_i(\mathbf{x},0) = 0, \quad \dot{u}_i(\mathbf{x},0) = 0, \quad p(\mathbf{x},0) = 0, \quad q(\mathbf{x},0) = 0, \qquad \mathbf{x} \in \Omega. \tag{5.42}$$

Since Ω is an unbounded spatial domain we impose the (weak) spatial growth restrictions

$$|u_{i,j}|,\ |\dot{u}_i|,\ |p|,\ |q|,\ |p_{,i}|,\ |q_{,i}| \leq K e^{\lambda r} \qquad \text{as} \qquad r \to \infty \tag{5.43}$$

where K and λ are positive constants. We also impose the following bounds on the coefficients in (5.40),

$$|a_{ijkh}|,\ |\beta_{ij}|,\ |\gamma_{ij}|,\ |k_{ij}|,\ |m_{ij}| \leq M \tag{5.44}$$

and

$$\begin{aligned} &a_{ijkh} \xi_{ij} \xi_{kh} \geq \xi_0 \xi_{ij} \xi_{ij} \qquad \text{for all} \quad \xi_{ij} \\ &k_{ij} \xi_i \xi_j \geq k_0 \xi_i \xi_i \qquad \text{for all} \quad \xi_i \\ &m_{ij} \xi_i \xi_j \geq m_0 \xi_i \xi_i \qquad \text{for all} \quad \xi_i \\ &\rho \geq \rho_0 > 0, \qquad \alpha \geq \alpha_0 > 0, \qquad \beta \geq \beta_0 > 0 \end{aligned} \tag{5.45}$$

where ξ_0, k_0, m_0 are positive constants.

The key to proving uniqueness of a solution to $\mathscr{P}$ by a weighted energy method is to follow section 1.3.5 and introduce a suitable weight, see Rionero & Galdi [178]. In this work we employ the weight function $g(r)$, $r=\sqrt{x_ix_i}$, as

$$g(r)=e^{-\delta r} \tag{5.46}$$

for $\delta>0$ a constant to be selected. More sophisticated choices of weight may be found in Galdi & Rionero [88]. Assume the solution to $\mathscr{P}$ satisfies the growth restriction (5.43) and we adopt conditions (5.44) and (5.45) on the coefficients.

The proof of uniqueness commences by multiplying equation $(5.40)_1$ by $g\dot{u}_i$ and integrating over Ω, assuming $\delta>2\lambda$. Observe that $g_{,i}=-\delta g x_i/r$, and then we may show

$$\begin{aligned}\frac{d}{dt}\frac{1}{2}\Big(\int_\Omega g\rho\dot{u}_i\dot{u}_i dx+\int_\Omega g a_{ijkh}u_{i,j}u_{k,h}dx\Big)&\\ =\int_\Gamma n_j g a_{ijkh}u_{k,h}\dot{u}_i dS+\int_\Gamma g\beta_{ij}p\dot{u}_i n_j dS+\int_\Gamma g\gamma_{ij}q\dot{u}_i n_j dS&\\ -\delta\int_\Omega g\frac{x_j}{r}a_{ijkh}u_{k,h}\dot{u}_i dx+\int_\Omega g\beta_{ij}p\dot{u}_{i,j}dx+\int_\Omega g\gamma_{ij}q\dot{u}_{i,j}dx&\\ -\delta\int_\Omega g\frac{x_j}{r}\beta_{ij}p\dot{u}_i dx-\delta\int_\Omega g\frac{x_j}{r}\gamma_{ij}q\dot{u}_i dx.&\end{aligned}$$

The boundary terms on Γ are zero due to the boundary conditions. We now use (5.44) and the Cauchy-Schwarz inequality to see that

$$\begin{aligned}\frac{d}{dt}\frac{1}{2}\Big(\int_\Omega g\rho\dot{u}_i\dot{u}_i dx+\int_\Omega g a_{ijkh}u_{i,j}u_{k,h}dx\Big)&\\ \leq\int_\Omega g\beta_{ij}p\dot{u}_{i,j}dx+\int_\Omega g\gamma_{ij}q\dot{u}_{i,j}dx&\\ +\delta M\int_\Omega g|\dot{u}_i|\,|u_{k,h}|\,dx+\delta M\int_\Omega g|\dot{u}_i|\,|p|\,dx&\\ +\delta M\int_\Omega g|\dot{u}_i|\,|q|\,dx.&\end{aligned} \tag{5.47}$$

Now multiply equation $(5.40)_2$ by gp and integrate over Ω, likewise multiply equation $(5.40)_3$ by gq and integrate over Ω, and then one may obtain

$$\begin{aligned}\frac{d}{dt}\frac{1}{2}\Big(\int_\Omega g\alpha p^2 dx\Big)+\int_\Omega g k_{ij}p_{,i}p_{,j}dx&\\ +\gamma\int_\Omega gp(p-q)dx=-\int_\Omega g\beta_{ij}p\dot{u}_{i,j}dx&\\ +\delta\int_\Omega g k_{ij}p_{,j}p\frac{x_i}{r}dx,&\end{aligned} \tag{5.48}$$

and

$$
\begin{aligned}
\frac{d}{dt}\frac{1}{2}\Big(\int_\Omega g\beta q^2 dx\Big) + \int_\Omega g m_{ij} q_{,i} q_{,j} dx & \\
-\gamma \int_\Omega g q(p-q) dx = - \int_\Omega g \gamma_{ij} q \dot{u}_{i,j} dx & \\
+\delta \int_\Omega g m_{ij} q_{,j} q \frac{x_i}{r} dx. &
\end{aligned}
\tag{5.49}
$$

The next step is to add (5.47) to (5.48) together with (5.49). We also employ a weighted arithmetic-geometric mean inequality and the bounds (5.45) to derive

$$
\begin{aligned}
\frac{dE}{dt} + \int_\Omega g k_{ij} p_{,i} p_{,j} dx + \int_\Omega g m_{ij} q_{,i} q_{,j} dx + \gamma \int_\Omega g (p-q)^2 dx & \\
\leq \frac{3M\delta}{2\rho_0} \int_\Omega g\rho \dot{u}_i \dot{u}_i dx + \frac{\delta M}{2\xi_0} \int_\Omega g a_{ijkh} u_{i,j} u_{k,h} dx & \\
+ \frac{\delta M}{\mu_1 \alpha_0} \int_\Omega g\alpha p^2 dx + \frac{\delta M}{\mu_2 \beta_0} \int_\Omega g\beta q^2 dx & \\
+ \frac{\delta M \mu_1}{2k_0} \int_\Omega g k_{ij} p_{,i} p_{,j} dx + \frac{\delta M \mu_2}{2m_0} \int_\Omega g m_{ij} q_{,i} q_{,j} dx, &
\end{aligned}
\tag{5.50}
$$

where $\mu_1 > 0$, $\mu_2 > 0$ are constants at our disposal and where the weighted energy $E(t)$ is defined by

$$
\begin{aligned}
E(t) = & \frac{1}{2} \int_\Omega g\rho \dot{u}_i \dot{u}_i dx + \frac{1}{2} \int_\Omega g a_{ijkh} u_{i,j} u_{k,h} dx \\
& + \frac{1}{2} \int_\Omega g\alpha p^2 dx + \frac{1}{2} \int_\Omega g\beta q^2 dx.
\end{aligned}
\tag{5.51}
$$

Now select $\mu_1 = 2k_0/\delta M$ and $\mu_2 = 2m_0/\delta M$. This removes the $p_{,i}$ and $q_{,i}$ terms on the right of (5.50). We further choose m to be the number

$$
m = \max\Big\{\frac{3M\delta}{2\rho_0}, \frac{\delta M}{2\xi_0}, \frac{\delta^2 M^2}{2k_0\alpha_0}, \frac{\delta^2 M^2}{2m_0\beta_0}\Big\}.
$$

Then, from inequality (5.50) we may deduce that

$$
\frac{dE}{dt} \leq mE. \tag{5.52}
$$

Integrate (5.52) from 0 to t to find

$$
0 \leq E(t) \leq e^{mt} E(0) = 0.
$$

Thus, given the definition of E, it follows that $p \equiv 0, q \equiv 0$ on Ω for $t > 0$, and $\nabla \mathbf{u} \equiv 0$ on the same domain. Since $\nabla \mathbf{u} \equiv 0$ on all of Ω we must have $u_i \equiv 0$ on Ω for all t, due to the boundary conditions. Uniqueness of the solution to $\mathscr{P}$ thus follows.

5.4 Exercises

Exercise 5.1. Consider the boundary-initial value problem $\mathscr{P}$ for the equations of a thermoelastic material with a double porosity structure, equations (2.34), but where Ω is an exterior domain with inner boundary Γ as in section 1.3.5. Thus $\mathscr{P}$ consists of the governing equations

$$\begin{aligned} \rho\ddot{u}_i &= (a_{ijkh}u_{k,h})_{,j} - (\beta_{ij}p)_{,j} - (\gamma_{ij}q)_{,j} - (a_{ij}\theta)_{,j} + \rho f_i\,, \\ \alpha\dot{p} &= (k_{ij}p_{,j})_{,i} - \gamma(p-q) - \beta_{ij}\dot{u}_{i,j} + \rho s_1\,, \\ \beta\dot{q} &= (m_{ij}q_{,j})_{,i} + \gamma(p-q) - \gamma_{ij}\dot{u}_{i,j} + \rho s_2\,, \\ a\dot{\theta} + a_{ij}\dot{u}_{i,j} &= (\kappa_{ij}\theta_{,j})_{,i} + \rho r. \end{aligned} \tag{5.53}$$

in $\Omega\times(0,T)$, with the boundary conditions

$$\begin{aligned} u_i(\mathbf{x},t) &= u_i^B(\mathbf{x},t), \quad p(\mathbf{x},t) = p^B(\mathbf{x},t), \\ q(\mathbf{x},t) &= q^B(\mathbf{x},t), \quad \theta(\mathbf{x},t) = \theta^B(\mathbf{x},t), \end{aligned} \tag{5.54}$$

on $\Gamma\times(0,T)$, and the initial conditions

$$\begin{aligned} &u_i(\mathbf{x},0) = v_i(\mathbf{x}), \quad \dot{u}_i(\mathbf{x},0) = w_i(\mathbf{x}), \quad p(\mathbf{x},0) = P(\mathbf{x}), \\ &q(\mathbf{x},0) = Q(\mathbf{x}), \quad \theta(\mathbf{x},0) = R(\mathbf{x}), \end{aligned} \tag{5.55}$$

$\mathbf{x}\in\Omega$.

Suppose a_{ijkh} satisfy the symmetries

$$a_{ijkh} = a_{jikh} = a_{khij}, \tag{5.56}$$

$\beta_{ij}, \gamma_{ij}, a_{ij}, k_{ij}, m_{ij}$ and κ_{ij} are symmetric and

$$\begin{aligned} &a_{ijkh}\xi_{ij}\xi_{kh} \geq a_0\xi_{ij}\xi_{ij}, \quad k_{ij}\xi_i\xi_j \geq k_0\xi_i\xi_i, \\ &m_{ij}\xi_i\xi_j \geq m_0\xi_i\xi_i, \quad \kappa_{ij}\xi_i\xi_j \geq \kappa_0\xi_i\xi_i, \end{aligned}$$

for all ξ_{ij}, ξ_i, where $a_0 > 0, k_0 > 0, m_0 > 0$ and $\kappa_0 > 0$ are constants. Suppose also

$$|a_{ijkh}|, |a_{ij}|, |\beta_{ij}|, |\gamma_{ij}| \leq M$$

for some constant $M < \infty$ and

$$|\dot{u}_i|, |u_{i,j}|, |p|, |p_{,i}|, |q|, |q_{,i}|, |\theta|, |\theta_{,i}| \leq ke^{\lambda r}$$

for some constants $k, \lambda > 0$ on $\Omega\times(0,T)$. By using a weighted energy method as in section 1.3.5 show that a solution to $\mathscr{P}$ is unique.

Exercise 5.2. By using a similar energy method to those of sections 5.2.1 and 5.2.2 show that the solution to the displacement boundary-initial value problem for equations (5.1) depends continuously on the source functions s_1 or s_2 when Ω is a bounded domain.

Exercise 5.3. Consider the double voids system of equations (3.69) for an isothermal elastic body on a bounded domain Ω. Let $\mathscr{P}$ be the boundary-initial value problem for those equations so that a solution (u_i, ν, ω) to $\mathscr{P}$ satisfies the system

$$\begin{aligned}
\rho \ddot{u}_i &= (a_{ijkh}u_{k,h})_{,j} + (b_{ij}\nu)_{,j} + (d_{ij}\omega)_{,j} + \rho f_i\,,\\
\kappa_1 \ddot{\nu} &= (\alpha_{ij}\nu_{,j})_{,i} + (\beta_{ij}\omega_{,j})_{,i} - \alpha_1 \nu - \alpha_3 \omega - b_{ij}u_{i,j} + \rho \ell_1\,,\\
\kappa_2 \ddot{\nu} &= (\gamma_{ij}\omega_{,j})_{,i} + (\beta_{ij}\nu_{,j})_{,i} - \alpha_2 \omega - \alpha_3 \nu - d_{ij}u_{i,j} + \rho \ell_2\,,
\end{aligned} \tag{5.57}$$

in $\Omega \times (0,T)$, together with the boundary conditions

$$\begin{aligned}
&u_i(\mathbf{x},t) = u_i^B(\mathbf{x},t), \quad \nu(\mathbf{x},t) = \nu^B(\mathbf{x},t),\\
&\omega(\mathbf{x},t) = \omega^B(\mathbf{x},t),
\end{aligned} \tag{5.58}$$

on $\Gamma \times (0,T)$, and the initial conditions

$$\begin{aligned}
&u_i(\mathbf{x},0) = v_i(\mathbf{x}), \quad \dot{u}_i(\mathbf{x},0) = w_i(\mathbf{x}),\\
&\nu(\mathbf{x},0) = \phi_1(\mathbf{x}), \quad \dot{\nu}(\mathbf{x},0) = \phi_2(\mathbf{x}),\\
&\omega(\mathbf{x},0) = \psi_1(\mathbf{x}), \quad \dot{\omega}(\mathbf{x},0) = \psi_2(\mathbf{x}),
\end{aligned} \tag{5.59}$$

$\mathbf{x} \in \Omega$.

Suppose a_{ijkh} satisfy the symmetries (5.56) and a_{ijkh} are positive-definite. Derive sufficient conditions on the symmetric tensors α_{ij}, β_{ij} and γ_{ij}, and upon the other coefficients to demonstrate that the solution to $\mathscr{P}$ depends continuously upon the initial data, and upon the body force f_i and the source terms ℓ_1 and ℓ_2. Use an energy method along the lines of sections 5.2.1 and 5.2.2.

Exercise 5.4. Consider the following set of equations for an elastic body with a double voids system, cf. equations (3.69), but where a Kelvin-Voigt viscoelastic effect is incorporated as well as viscoelastic effects in the voids distribution equations,

$$\begin{aligned}
\rho \ddot{u}_i &= (a_{ijkh}u_{k,h})_{,j} + (b_{ijkh}\dot{u}_{k,h})_{,j} + (b_{ij}\nu)_{,j} + (d_{ij}\omega)_{,j}\,,\\
\kappa_1 \ddot{\nu} &= (\alpha_{ij}\nu_{,j})_{,i} + (\beta_{ij}\omega_{,j})_{,i} - \alpha_1 \nu - \alpha_3 \omega - b_{ij}u_{i,j} - \tau_1 \dot{\nu}\,,\\
\kappa_2 \ddot{\nu} &= (\gamma_{ij}\omega_{,j})_{,i} + (\beta_{ij}\nu_{,j})_{,i} - \alpha_2 \omega - \alpha_3 \nu - d_{ij}u_{i,j} - \tau_2 \dot{\omega}\,.
\end{aligned} \tag{5.60}$$

Let $\mathscr{P}$ denote the boundary-initial value problem comprised of equations (5.60) in $\Omega \times \{t > 0\}$ together with the boundary conditions

$$u_i(\mathbf{x},t) = 0, \quad \nu(\mathbf{x},t) = 0, \quad \omega(\mathbf{x},t) = 0, \qquad \text{on } \Gamma \times \{t > 0\}, \tag{5.61}$$

and the initial conditions

$$\begin{aligned} u_i(\mathbf{x},0) &= v_i(\mathbf{x}), \quad \dot{u}_i(\mathbf{x},0) = w_i(\mathbf{x}), \\ v(\mathbf{x},0) &= \phi_1(\mathbf{x}), \quad \dot{v}(\mathbf{x},0) = \phi_2(\mathbf{x}), \\ \omega(\mathbf{x},0) &= \psi_1(\mathbf{x}), \quad \dot{\omega}(\mathbf{x},0) = \psi_2(\mathbf{x}), \end{aligned} \tag{5.62}$$

$\mathbf{x} \in \Omega$.

Suppose a_{ijkh} and b_{ijkh} satisfy the symmetries (5.56) with the other tensors being symmetric. Suppose also $\alpha_1, \alpha_2, \tau_1, \tau_2\ \rho, \kappa_1, \kappa_2$ are strictly positive and

$$\alpha_{ij}\xi_i\xi_j + \gamma_{ij}\mu_i\mu_j + 2\beta_{ij}\xi_i\mu_j \geq 0, \qquad \forall\, \xi_i, \mu_j$$

with

$$a_{ijkh}\xi_{ij}\xi_{kh} \geq a_0\xi_{ij}\xi_{ij}, \quad b_{ijkh}\xi_{ij}\xi_{kh} \geq b_0\xi_{ij}\xi_{ij}, \qquad \forall\, \xi_{ij}, \quad \text{some } a_0 > 0, b_0 > 0.$$

In addition suppose

$$\begin{aligned} &\rho_m \geq \rho(\mathbf{x}) \geq \rho_0 > 0, \\ &\tau_1^{max} \geq \tau_1(\mathbf{x}) \geq \tau_1^{min}, \qquad \tau_2^{max} \geq \tau_2(\mathbf{x}) \geq \tau_2^{min}, \\ &\kappa_1^{max} \geq \kappa_1(\mathbf{x}) \geq \kappa_1^{min}, \qquad \kappa_2^{max} \geq \kappa_2(\mathbf{x}) \geq \kappa_2^{min}. \end{aligned}$$

By using the transformation technique of exercise 1.12, show that a solution to $\mathscr{P}$ will decay exponentially in $< \rho u_i u_i >, < \kappa_1 v^2 >, < \kappa_2 \omega^2 >$ measure at least like $e^{-2\lambda t}$, for some constant λ, such that

$$0 < \lambda \leq \min\left\{ \frac{\tau_1^{min}}{\kappa_1^{max}}, \frac{\tau_2^{min}}{\kappa_2^{max}}, \frac{b_0\lambda_1}{2\rho_m} \right\}$$

where λ_1 is the constant in the Poincaré inequality for Ω, provided the coefficients satisfy the relation

$$\begin{aligned} &< a_{ijkh}u_{i,j}u_{k,h} > +2 < b_{ij}u_{i,j}v > +2 < d_{ij}u_{i,j}\omega > + < \alpha_1 v^2 > + < \alpha_2\omega^2 > \\ &\quad +2 < \alpha_3\omega v > \geq \lambda^*\big[< b_{ijkh}u_{i,j}u_{k,h} > + < \tau_1 v^2 > + < \tau_2\omega^2 > \big]. \end{aligned}$$

Chapter 6
Uniqueness Without Definiteness Conditions

In this chapter the elastic coefficients a_{ijkh} have the condition of positive definiteness or even positivity relaxed and all we require is that they satisfy the symmetry conditions (1.16). We have already given cogent reasons in the first paragraph of section 1.4.4 why it is important to consider the situation when only symmetry is stipulated and no definiteness is demanded.

Our goal in this chapter is to establish uniqueness of a solution to three classes of multi - porosity elasticity problems. We shall not impose any definiteness conditions on the elastic coefficients and instead require only the symmetries of (1.16) which we repeat here for clarity,

$$a_{ijkh} = a_{jikh} = a_{khij} \,. \tag{6.1}$$

The three classes of elastic body we consider are double porosity thermoelasticity, isothermal triple porosity elasticity, and triple porosity thermoelasticity. The governing equations for these three materials may be found in (2.34), (2.44) and (2.47), respectively. While we employ a logarithmic convexity method for each class of problem the functional to be used is different for each case. In some sense, the different functions arise naturally for each problem, and we believe it is interesting from a mathematical viewpoint to analyse the various methods.

6.1 Double Porosity Thermoelasticity

We commence with the displacement boundary-initial value problem for double porosity thermoelasticity governed by equations (2.34). Observe that there are no cross inertia terms in (2.34) involving p,q and θ. One can establish uniqueness for the problem which involves cross inertia terms and details may be found in Straughan [209].

Let $(u_i^1, p^1, q^1, \theta^1)$ and $(u_i^2, p^2, q^2, \theta^2)$ be solutions to equations (2.34) which both satisfy the boundary conditions

B. Straughan, *Mathematical Aspects of Multi–Porosity Continua*, Advances in Mechanics and Mathematics 38, https://doi.org/10.1007/978-3-319-70172-1_6

$$
\begin{aligned}
&u_i(\mathbf{x},t) = u_i^B(\mathbf{x},t), \qquad p(\mathbf{x},t) = p^B(\mathbf{x},t), \\
&q(\mathbf{x},t) = q^B(\mathbf{x},t), \qquad \theta(\mathbf{x},t) = \theta^B(\mathbf{x},t), \qquad \text{on } \Gamma \times (0,T),
\end{aligned} \tag{6.2}
$$

and the initial conditions

$$
\begin{aligned}
&u_i(\mathbf{x},0) = v_i(\mathbf{x}), \qquad \dot{u}_i(\mathbf{x},0) = w_i(\mathbf{x}), \qquad p(\mathbf{x},0) = P(\mathbf{x}), \\
&q(\mathbf{x},0) = Q(\mathbf{x}), \qquad \theta(\mathbf{x},0) = R(\mathbf{x}), \qquad \mathbf{x} \in \Omega,
\end{aligned} \tag{6.3}
$$

for the same given functions $u_i^B, p^B, q^B, \theta^B, v_i, w_i, P, Q$ and R, and for the same supply functions f_i, s_1, s_2 and r.

Define the difference solution (u_i, p, q, θ) by

$$
u_i = u_i^1 - u_i^2, \quad p = p^1 - p^2, \quad q = q^1 - q^2, \quad \theta = \theta^1 - \theta^2, \tag{6.4}
$$

and then one observes that this solution satisfies the boundary - initial value problem

$$
\begin{aligned}
&\rho \ddot{u}_i = (a_{ijkh} u_{k,h})_{,j} - (\beta_{ij} p)_{,j} - (\gamma_{ij} q)_{,j} - (a_{ij} \theta)_{,j}, \\
&\alpha \dot{p} = (k_{ij} p_{,j})_{,i} - \gamma(p-q) - \beta_{ij} \dot{u}_{i,j}, \\
&\beta \dot{q} = (m_{ij} q_{,j})_{,i} + \gamma(p-q) - \gamma_{ij} \dot{u}_{i,j}, \\
&a \dot{\theta} = (\kappa_{ij} \theta_{,j})_{,i} - a_{ij} \dot{u}_{i,j},
\end{aligned} \tag{6.5}
$$

together with

$$
\begin{aligned}
&u_i(\mathbf{x},t) = 0, \qquad p(\mathbf{x},t) = 0, \\
&q(\mathbf{x},t) = 0, \qquad \theta(\mathbf{x},t) = 0, \qquad \text{on } \Gamma \times (0,T),
\end{aligned} \tag{6.6}
$$

and

$$
\begin{aligned}
&u_i(\mathbf{x},0) = 0, \qquad \dot{u}_i(\mathbf{x},0) = 0, \qquad p(\mathbf{x},0) = 0, \\
&q(\mathbf{x},0) = 0, \qquad \theta(\mathbf{x},0) = 0, \qquad \mathbf{x} \in \Omega.
\end{aligned} \tag{6.7}
$$

The functions $\beta_{ij}, \gamma_{ij}, a_{ij}, k_{ij}, m_{ij}$ and κ_{ij} are symmetric and $k_{ij}, m_{ij}, \kappa_{ij}$ are non-negative. The coefficients ρ, α, β and a are strictly positive and $\gamma \geq 0$.

To demonstrate uniqueness of a solution to (2.34) with conditions (6.2) and (6.3) we employ a logarithmic convexity technique. Firstly we calculate the energy equation associated with (6.5) - (6.7). Multiply equations $(6.5)_1$ - $(6.5)_4$ in turn by $\dot{u}_i, p, q$ and then θ and integrate each over Ω. After some integration by parts, and use of (6.6) and an integration in time, one may derive the energy equation

$$
\begin{aligned}
E(t) &+ \int_0^t < k_{ij} p_{,i} p_{,j} > ds + \int_0^t < m_{ij} q_{,i} q_{,j} > ds \\
&+ \int_0^t < \kappa_{ij} \theta_{,i} \theta_{,j} > ds + \int_0^t < \gamma(p-q)^2 > ds = E(0) = 0,
\end{aligned} \tag{6.8}
$$

where

$$E(t) =\frac{1}{2} < \rho\dot{u}_i\dot{u}_i > +\frac{1}{2} < a_{ijkh}u_{i,j}u_{k,h} > +\frac{1}{2} < \alpha p^2 > \\ +\frac{1}{2} < \beta q^2 > +\frac{1}{2} < a\theta^2 >, \tag{6.9}$$

and where $E(0) = 0$ due to (6.7).

Next, introduce the variables η, ζ and ψ by

$$\eta(\mathbf{x},t) = \int_0^t p(\mathbf{x},s)ds, \quad \zeta(\mathbf{x},t) = \int_0^t q(\mathbf{x},s)ds, \quad \psi(\mathbf{x},t) = \int_0^t \theta(\mathbf{x},s)ds. \tag{6.10}$$

Then we introduce what appears to be a natural function for the method of logarithmic convexity, namely $F(t)$, where

$$F(t) = < \rho u_i u_i > + \int_0^t < k_{ij}\eta_{,i}\eta_{,j} > ds + \int_0^t < m_{ij}\zeta_{,i}\zeta_{,j} > ds \\ + \int_0^t < \kappa_{ij}\psi_{,i}\psi_{,j} > ds + \int_0^t < \gamma(\eta-\zeta)^2 > ds. \tag{6.11}$$

By differentiation and use of the initial conditions one now shows

$$F'(t) =2 < \rho u_i\dot{u}_i > +2\int_0^t < k_{ij}\eta_{,i}p_{,j} > ds + 2\int_0^t < m_{ij}\zeta_{,i}q_{,j} > ds \\ +2\int_0^t < \kappa_{ij}\psi_{,i}\theta_{,j} > ds + 2\int_0^t < \gamma(\eta-\zeta)(p-q) > ds. \tag{6.12}$$

We now differentiate again to find

$$F'' =2 < \rho\dot{u}_i\dot{u}_i > +2 < \rho u_i\ddot{u}_i > +2 < k_{ij}\eta_{,i}p_{,j} > +2 < m_{ij}\zeta_{,i}q_{,j} > \\ +2 < \kappa_{ij}\psi_{,i}\theta_{,j} > +2 < \gamma(\eta-\zeta)(p-q) >,$$

and when we substitute for $\rho\ddot{u}_i$ from $(6.5)_1$ we find after some integration by parts

$$F'' =2 < \rho\dot{u}_i\dot{u}_i > -2 < a_{ijkh}u_{i,j}u_{k,h} > +2 < p\beta_{ij}u_{i,j} > +2 < q\gamma_{ij}u_{i,j} > \\ +2 < a_{ij}\theta u_{i,j} > +2 < \gamma(\eta-\zeta)(p-q) > +2 < k_{ij}\eta_{,i}p_{,j} > \\ +2 < m_{ij}\zeta_{,i}q_{,j} > +2 < \kappa_{ij}\psi_{,i}\theta_{,j} > . \tag{6.13}$$

We now integrate equations $(6.5)_{2-4}$ in time and use (6.7) to see that

$$\alpha p =(k_{ij}\eta_{,j})_{,i} - \gamma(\eta-\zeta) - \beta_{ij}u_{i,j}, \\ \beta q =(m_{ij}\zeta_{,j})_{,i} + \gamma(\eta-\zeta) - \gamma_{ij}u_{i,j}, \\ a\theta =(\kappa_{ij}\psi_{,j})_{,i} - a_{ij}u_{i,j}. \tag{6.14}$$

Multiply $(6.14)_1$ by p, $(6.14)_2$ by q, $(6.14)_3$ by θ, integrate each over Ω using the boundary conditions, and form the sum of the three resulting equations to find

$$
\begin{aligned}
&2 < p\beta_{ij}u_{i,j} > +2 < q\gamma_{ij}u_{i,j} > +2 < a_{ij}\theta u_{i,j} > +2 < \gamma(p-q)(\eta-\zeta) > \\
&\quad +2 < k_{ij}\eta_{,j}p_{,i} > +2 < m_{ij}\zeta_{,j}q_{,i} > +2 < \kappa_{ij}\psi_{,i}\theta_{,j} > \\
&\quad = -2 < \alpha p^2 > -2 < \beta q^2 > -2 < a\theta^2 > .
\end{aligned} \tag{6.15}
$$

Substitute now from (6.15) in (6.13) to obtain

$$
F'' = 2 < \rho\dot{u}_i\dot{u}_i > -2 < a_{ijkh}u_{i,j}u_{k,h} > -2 < \alpha p^2 > -2 < \beta q^2 > -2 < a\theta^2 >,
$$

and then substitute for the last four terms using (6.9) and (6.8) to obtain

$$
\begin{aligned}
F'' =& 4 < \rho\dot{u}_i\dot{u}_i > +4\int_0^t < k_{ij}p_{,i}p_{,j} > ds + 4\int_0^t < m_{ij}q_{,i}q_{,j} > ds \\
&+4\int_0^t < \kappa_{ij}\theta_{,i}\theta_{,j} > ds + 4\int_0^t < \gamma(p-q)^2 > ds.
\end{aligned} \tag{6.16}
$$

With the aid of (6.11), (6.12) and (6.16) one may then assert that

$$
FF'' - (F')^2 = 4S^2, \tag{6.17}
$$

where

$$
\begin{aligned}
S^2 =& \Big[< \rho u_i u_i > + \int_0^t < k_{ij}\eta_{,i}\eta_{,j} > ds + \int_0^t < m_{ij}\zeta_{,i}\zeta_{,j} > ds \\
&+ \int_0^t < \kappa_{ij}\psi_{,i}\psi_{,j} > ds + \int_0^t < \gamma(\eta-\zeta)^2 > ds \Big] \\
&\times \Big[< \rho\dot{u}_i\dot{u}_i > + \int_0^t < k_{ij}p_{,i}p_{,j} > ds + \int_0^t < m_{ij}q_{,i}q_{,j} > ds \\
&+ \int_0^t < \kappa_{ij}\theta_{,i}\theta_{,j} > ds + \int_0^t < \gamma(p-q)^2 > ds \Big] \\
&- \Big[< \rho u_i\dot{u}_i > + \int_0^t < k_{ij}\eta_{,i}p_{,j} > ds + \int_0^t < m_{ij}\zeta_{,i}q_{,j} > ds \\
&+ \int_0^t < \kappa_{ij}\psi_{,i}\theta_{,j} > ds + \int_0^t < \gamma(\eta-\zeta)(p-q) > ds \Big]^2 .
\end{aligned} \tag{6.18}
$$

From (6.17) it follows that $F \equiv 0$ on $t \in (0,T)$, see section 1.3.6. Thus, $u_i \equiv 0$ on $\Omega \times [0,T]$. Then from the energy equation (6.8) one finds

$$
0 \leq < \alpha p^2 > + < \beta q^2 > + < a\theta^2 > \leq 0, \qquad t \in (0,T),
$$

and hence $p \equiv 0, q \equiv 0, \theta \equiv 0$. Thus, a solution to the double porosity thermoelasticity problem is unique.

Uniqueness and growth results for a thermoelastic body with dipolar structure and with a double porosity are given by Marin & Nicaise [157].

6.2 Triple Porosity Elasticity

In this section we investigate the uniqueness question for the displacement boundary - initial value problem for the isothermal triple porosity equations (2.44) assuming only the symmetry conditions (6.1) on the elastic coefficients and no definiteness.

Let $\mathscr{P}$ denote the boundary - initial value problem for equations (2.44) subject to thc boundary conditions

$$\begin{aligned} &u_i(\mathbf{x},t) = u_i^B(\mathbf{x},t), \qquad && p(\mathbf{x},t) = p^B(\mathbf{x},t), \\ &q(\mathbf{x},t) = q^B(\mathbf{x},t), && s(\mathbf{x},t) = s^B(\mathbf{x},t), \qquad \text{on } \Gamma \times (0,T), \end{aligned} \tag{6.19}$$

and the initial conditions

$$\begin{aligned} &u_i(\mathbf{x},0) = v_i(\mathbf{x}), \qquad && \dot{u}_i(\mathbf{x},0) = w_i(\mathbf{x}), \qquad && p(\mathbf{x},0) = P(\mathbf{x}), \\ &q(\mathbf{x},0) = Q(\mathbf{x}), && s(\mathbf{x},0) = S(\mathbf{x}), && \mathbf{x} \in \Omega. \end{aligned} \tag{6.20}$$

Let (u_i^1, p^1, q^1, s^1) and (u_i^2, p^2, q^2, s^2) be two solutions to $\mathscr{P}$ for the same boundary and initial data (6.19) and (6.20) and for the same supply functions f_i, s_1, s_2 and s_3. Define the difference solution (u_i, p, q, s) by

$$u_i = u_i^1 - u_i^2, \quad p = p^1 - p^2, \quad q = q^1 - q^2, \quad s = s^1 - s^2,$$

and then one verfies that this solution satisfies the boundary - initial value problem

$$\begin{aligned} &\rho \ddot{u}_i = (a_{ijkh} u_{k,h})_{,j} - (\beta_{ij} p)_{,j} - (\gamma_{ij} q)_{,j} - (\omega_{ij} s)_{,j}, \\ &\alpha \dot{p} = (k_{ij} p_{,j})_{,i} - \gamma(p - q) - \beta_{ij} \dot{u}_{i,j}, \\ &\beta \dot{q} = (m_{ij} q_{,j})_{,i} + \gamma(p - q) + \xi(s - q) - \gamma_{ij} \dot{u}_{i,j}, \\ &\varepsilon \dot{s} = (\ell_{ij} s_{,j})_{,i} - \xi(s - q) - \omega_{ij} \dot{u}_{i,j}, \end{aligned} \tag{6.21}$$

alongside the boundary conditions

$$\begin{aligned} &u_i(\mathbf{x},t) = 0, \qquad && p(\mathbf{x},t) = 0, \\ &q(\mathbf{x},t) = 0, && s(\mathbf{x},t) = 0, \qquad \text{on } \Gamma \times (0,T), \end{aligned} \tag{6.22}$$

and the initial conditions

$$\begin{aligned} &u_i(\mathbf{x},0) = 0, \qquad && \dot{u}_i(\mathbf{x},0) = 0, \qquad && p(\mathbf{x},0) = 0, \\ &q(\mathbf{x},0) = 0, && s(\mathbf{x},0) = 0, && \mathbf{x} \in \Omega. \end{aligned} \tag{6.23}$$

The functions $\beta_{ij}, \gamma_{ij}, \omega_{ij}, k_{ij}, m_{ij}$ and ℓ_{ij} are symmetric and $k_{ij}, m_{ij}, \ell_{ij}$ are non-negative. The terms $\rho, \alpha, \beta, \varepsilon$ are strictly positive with $\gamma \geq 0$ and $\xi \geq 0$.

To demonstrate uniqueness of a solution to $\mathscr{P}$ we again employ a logarithmic convexity technique, but necessarily the key function in the proof is different. To commence we calculate the energy equation associated with (6.21) - (6.23). One now multiplies equations $(6.21)_1$ - $(6.21)_4$ respectively by $\dot{u}_i, p, q$ and then s and one integrates each resulting equation over Ω. One employs some integration by parts, one uses the boundary conditions (6.22) and then one integrates in time. In this manner one may derive the energy equation

$$\begin{aligned} E(t) &+ \int_0^t < k_{ij}p_{,i}p_{,j} > ds + \int_0^t < m_{ij}q_{,i}q_{,j} > ds \\ &+ \int_0^t < \ell_{ij}s_{,i}s_{,j} > ds + \int_0^t < \gamma(p-q)^2 > ds \\ &+ \int_0^t < \xi(s-q)^2 > ds = E(0) = 0, \end{aligned} \tag{6.24}$$

where the energy function is defined by

$$\begin{aligned} E(t) =& \frac{1}{2} < \rho \dot{u}_i \dot{u}_i > + \frac{1}{2} < a_{ijkh}u_{i,j}u_{k,h} > + \frac{1}{2} < \alpha p^2 > \\ &+ \frac{1}{2} < \beta q^2 > + \frac{1}{2} < \varepsilon s^2 > . \end{aligned} \tag{6.25}$$

Observe that $E(0) = 0$ because of the initial data (6.23).

Define the variables η, ζ and ϕ by

$$\eta(\mathbf{x},t) = \int_0^t p(\mathbf{x},a)da, \quad \zeta(\mathbf{x},t) = \int_0^t q(\mathbf{x},a)da, \quad \phi(\mathbf{x},t) = \int_0^t s(\mathbf{x},a)da, \tag{6.26}$$

and then from integration of equations $(6.21)_{2-4}$ one finds

$$\begin{aligned} \alpha p =& (k_{ij}\eta_{,j})_{,i} - \gamma(\eta - \zeta) - \beta_{ij}u_{i,j}, \\ \beta q =& (m_{ij}\zeta_{,j})_{,i} + \gamma(\eta - \zeta) + \xi(\phi - \zeta) - \gamma_{ij}u_{i,j}, \\ \varepsilon s =& (\ell_{ij}\phi_{,j})_{,i} - \xi(\phi - \zeta) - \omega_{ij}u_{i,j}. \end{aligned} \tag{6.27}$$

In this section we base a logarithmic convexity method on the function $G(t)$, where

$$\begin{aligned} G(t) = & < \rho u_i u_i > + \int_0^t < k_{ij}\eta_{,i}\eta_{,j} > da + \int_0^t < m_{ij}\zeta_{,i}\zeta_{,j} > da \\ &+ \int_0^t < \ell_{ij}\phi_{,i}\phi_{,j} > da + \int_0^t < \gamma(\eta - \zeta)^2 > da \\ &+ \int_0^t < \xi(\phi - \zeta)^2 > da. \end{aligned} \tag{6.28}$$

The procedure is to now differentiate the function G as given by expression (6.28). After employment of the definitions of η, ζ and ϕ given in equations (6.26), and utilizing the initial conditions (6.23), one sees that

$$\begin{aligned} G'(t) =& 2 < \rho u_i \dot{u}_i > + 2\int_0^t < k_{ij}\eta_{,i}p_{,j} > da + 2\int_0^t < m_{ij}\zeta_{,i}q_{,j} > da \\ &+ 2\int_0^t < \ell_{ij}\phi_{,i}s_{,j} > da + 2\int_0^t < \gamma(\eta-\zeta)(p-q) > da \\ &+ 2\int_0^t < \xi(\phi-\zeta)(s-q) > da. \end{aligned} \tag{6.29}$$

One now multiplies equation $(6.21)_1$ by u_i, equation $(6.27)_1$ by p, equation $(6.27)_2$ by q, equation $(6.27)_3$ by s, then integrates each over Ω to find

$$\begin{aligned} & < \rho \ddot{u}_i u_i > + < a_{ijkh}u_{i,j}u_{k,h} > - < \beta_{ij}pu_{i,j} > \\ & \quad - < \gamma_{ij}qu_{i,j} > - < \omega_{ij}su_{i,j} > = 0, \\ & < \alpha p^2 > + < k_{ij}\eta_{,j}p_{,i} > + < \gamma(\eta-\zeta)p > + < \beta_{ij}pu_{i,j} > = 0, \\ & < \beta q^2 > + < m_{ij}\zeta_{,j}q_{,i} > - < \gamma(\eta-\zeta)q > \\ & \quad - < \xi q(\phi-\zeta) > + < \gamma_{ij}qu_{i,j} > = 0, \\ & < \varepsilon s^2 > + < \ell_{ij}\phi_{,j}s_{,i} > + < \xi s(\phi-\zeta) > + < \omega_{ij}su_{i,j} > = 0. \end{aligned} \tag{6.30}$$

Upon adding these equations we may derive the useful relation

$$\begin{aligned} & < \rho \ddot{u}_i u_i > + < a_{ijkh}u_{i,j}u_{k,h} > + < \alpha p^2 > + < k_{ij}\eta_{,j}p_{,i} > \\ & + < \varepsilon s^2 > + < \beta q^2 > + < m_{ij}\zeta_{,j}q_{,i} > + < \ell_{ij}\phi_{,j}s_{,i} > \\ & + < \gamma(\eta-\zeta)(p-q) > + < \xi(s-q)(\phi-\zeta) > = 0. \end{aligned} \tag{6.31}$$

One then differentiates G' in (6.29) to obtain

$$\begin{aligned} G'' =& 2 < \rho \dot{u}_i \dot{u}_i > + 2 < \rho u_i \ddot{u}_i > + 2 < k_{ij}\eta_{,i}p_{,j} > + 2 < m_{ij}\zeta_{,i}q_{,j} > \\ &+ 2 < \ell_{ij}\phi_{,i}s_{,j} > + 2 < \gamma(\eta-\zeta)(p-q) > + 2 < \xi(\phi-\zeta)(s-q) > . \end{aligned}$$

Now, employ (6.31) in the expression for G'' to derive

$$G'' = 2 < \rho \dot{u}_i \dot{u}_i > - 2 < a_{ijkh}u_{i,j}u_{k,h} > - 2 < \alpha p^2 > - 2 < \beta q^2 > - 2 < \varepsilon s^2 > .$$

Finally we use the energy equation (6.24) to further rewrite G'' as

$$\begin{aligned} G'' =& 4 < \rho \dot{u}_i \dot{u}_i > + 4\int_0^t < k_{ij}p_{,i}p_{,j} > da + 4\int_0^t < m_{ij}q_{,i}q_{,j} > da \\ &+ 4\int_0^t < \ell_{ij}s_{,i}s_{,j} > da + 4\int_0^t < \gamma(p-q)^2 > da \\ &+ 4\int_0^t < \xi(s-q)^2 > da. \end{aligned} \tag{6.32}$$

Using (6.28), (6.29) and (6.32) we now form

$$GG'' - (G')^2 = 4S_1^2 \geq 0, \tag{6.33}$$

where S_1^2, which is non-negative thanks to the Cauchy-Schwarz inequality, is given by

$$S_1^2 = AB - C^2, \tag{6.34}$$

where A, B and C have the forms

$$A = <\rho u_i u_i> + \int_0^t <k_{ij}\eta_{,i}\eta_{,j}> da + \int_0^t <m_{ij}\zeta_{,i}\zeta_{,j}> da \\ + \int_0^t <\ell_{ij}\phi_{,i}\phi_{,j}> da + \int_0^t <\gamma(\eta-\zeta)^2> da + \int_0^t <\xi(\phi-\zeta)^2> da$$

and

$$B = <\rho \dot{u}_i \dot{u}_i> + \int_0^t <k_{ij}p_{,i}p_{,j}> da + \int_0^t <m_{ij}q_{,i}q_{,j}> da \\ + \int_0^t <\ell_{ij}s_{,i}s_{,j}> ds + \int_0^t <\gamma(p-q)^2> da + \int_0^t <\xi(s-q)^2> da$$

with

$$C = <\rho u_i \dot{u}_i> + \int_0^t <k_{ij}\eta_{,i}p_{,j}> da + \int_0^t <m_{ij}\zeta_{,i}q_{,j}> da \\ + \int_0^t <\ell_{ij}\phi_{,i}s_{,j}> da + \int_0^t <\gamma(\eta-\zeta)(p-q)> da \\ + \int_0^t <\xi(\phi-\zeta)(s-q)> da$$

From inequality (6.33) one may deduce $G \equiv 0$ on $t \in (0,T)$, see section 1.3.6, and then from (6.28) it follows that $u_i \equiv 0$ on $\Omega \times [0,T]$. Then, one may employ (6.25) and (6.24) to infer that

$$p \equiv 0, \qquad q \equiv 0, \qquad s \equiv 0, \qquad \text{on } \Omega \times [0,T].$$

Hence, uniqueness of a solution to $\mathscr{P}$ follows.

6.3 Triple Porosity Thermoelasticity

We proceed now to demonstrate uniqueness of a solution to the displacement boundary - initial value problem for equations (2.47) which govern the behaviour of a thermoelastic body with a triple porosity structure. We denote the appropriate boundary-initial value problem in this case by $\mathscr{T}$ and this thus consists of equations (2.47) for u_i, p, q, s and θ together with the boundary conditions

$$\begin{aligned} &u_i(\mathbf{x},t) = u_i^B(\mathbf{x},t), \qquad p(\mathbf{x},t) = p^B(\mathbf{x},t), \qquad q(\mathbf{x},t) = q^B(\mathbf{x},t), \\ &s(\mathbf{x},t) = s^B(\mathbf{x},t), \qquad \theta(\mathbf{x},t) = \theta^B(\mathbf{x},t), \qquad \text{on } \Gamma \times (0,T), \end{aligned} \tag{6.35}$$

and the initial conditions

$$\begin{aligned} &u_i(\mathbf{x},0)=v_i(\mathbf{x}), && \dot{u}_i(\mathbf{x},0)=w_i(\mathbf{x}), && p(\mathbf{x},0)=P(\mathbf{x}), \\ &q(\mathbf{x},0)=Q(\mathbf{x}), && s(\mathbf{x},0)=S(\mathbf{x}), && \theta(\mathbf{x},0)=R(\mathbf{x}), \qquad \mathbf{x}\in\Omega. \end{aligned} \tag{6.36}$$

Let $(u_i^1,p^1,q^1,s^1,\theta^1)$ and $(u_i^2,p^2,q^2,s^2,\theta^2)$ be two solutions which satisfy $\mathscr{T}$ for the same boundary data u_i^B,p^B,q^B,s^B and θ^B, for the same initial data v_i,w_i,P,Q,S and R, and for the same supply functions f_i,s_1,s_2,s_3 and r.

Define the difference solution (u_i,p,q,s,θ) by

$$\begin{aligned} &u_i=u_i^1-u_i^2, \quad p=p^1-p^2, \quad q=q^1-q^2, \\ &s=s^1-s^2, \qquad \theta=\theta^1-\theta^2. \end{aligned} \tag{6.37}$$

The difference solution is seen to satisfy the boundary - initial value problem

$$\begin{aligned} &\rho\ddot{u}_i=(a_{ijkh}u_{k,h})_{,j}-(\beta_{ij}p)_{,j}-(\gamma_{ij}q)_{,j}-(\omega_{ij}s)_{,j}-(a_{ij}\theta)_{,j}, \\ &\alpha\dot{p}=(k_{ij}p_{,j})_{,i}-\gamma(p-q)-\beta_{ij}\dot{u}_{i,j}, \\ &\beta\dot{q}=(m_{ij}q_{,j})_{,i}+\gamma(p-q)-\xi(q-s)-\gamma_{ij}\dot{u}_{i,j}, \\ &\varepsilon\dot{s}=(\ell_{ij}q_{,j})_{,i}+\xi(q-s)-\omega_{ij}\dot{u}_{i,j}, \\ &a\theta=(r_{ij}\theta_{,j})_{,i}-a_{ij}\dot{u}_{i,j}, \end{aligned} \tag{6.38}$$

together with boundary conditions

$$\begin{aligned} &u_i(\mathbf{x},t)=0, && p(\mathbf{x},t)=0, && q(\mathbf{x},t)=0, \\ &s(\mathbf{x},t)=0, && \theta(\mathbf{x},t)=0, && \text{on } \Gamma\times(0,T), \end{aligned} \tag{6.39}$$

and the initial conditions

$$\begin{aligned} &u_i(\mathbf{x},0)=0, && \dot{u}_i(\mathbf{x},0)=0, && p(\mathbf{x},0)=0, \\ &q(\mathbf{x},0)=0, && s(\mathbf{x},0)=0, && \theta(\mathbf{x},0)=0, \qquad \mathbf{x}\in\Omega. \end{aligned} \tag{6.40}$$

The functions $\beta_{ij},\gamma_{ij},\omega_{ij},a_{ij},k_{ij},m_{ij},\ell_{ij}$ and r_{ij} are symmetric and $k_{ij},m_{ij},\ell_{ij},r_{ij}$ are non-negative. The terms $\rho,\alpha,\beta,\varepsilon,a$ are strictly positive and $\gamma\geq 0,\xi\geq 0$.

For the problem of this section the energy function $E(t)$ is

$$\begin{aligned} E(t)=&\frac{1}{2}<\rho\dot{u}_i\dot{u}_i>+\frac{1}{2}<a_{ijkh}u_{i,j}u_{k,h}>+\frac{1}{2}<\alpha p^2> \\ &+\frac{1}{2}<\beta q^2>+\frac{1}{2}<\varepsilon s^2>+\frac{1}{2}<a\theta^2>. \end{aligned} \tag{6.41}$$

From (6.38) - (6.41) one may verify that the energy function satisfies the equation

$$
\begin{aligned}
E(t) &+ \int_0^t < k_{ij}p_{,i}p_{,j} > da + \int_0^t < m_{ij}q_{,i}q_{,j} > da \\
&+ \int_0^t < \ell_{ij}s_{,i}s_{,j} > da + \int_0^t < r_{ij}\theta_{,i}\theta_{,j} > da \\
&+ \int_0^t < \gamma(p-q)^2 > da + \int_0^t < \xi(s-q)^2 > da = E(0) = 0.
\end{aligned} \tag{6.42}
$$

We introduce the variables η, ζ, ϕ and ψ by

$$
\begin{aligned}
\eta(\mathbf{x},t) &= \int_0^t p(\mathbf{x},a)da, \quad \zeta(\mathbf{x},t) = \int_0^t q(\mathbf{x},a)da, \\
\phi(\mathbf{x},t) &= \int_0^t s(\mathbf{x},a)da, \quad \psi(\mathbf{x},t) = \int_0^t \theta(\mathbf{x},a)da.
\end{aligned} \tag{6.43}
$$

By integrating equations $(6.38)_{2-5}$ in t we derive the useful relations

$$
\begin{aligned}
\alpha p &= (k_{ij}\eta_{,j})_{,i} - \gamma(\eta-\zeta) - \beta_{ij}u_{i,j}, \\
\beta q &= (m_{ij}\zeta_{,j})_{,i} + \gamma(\eta-\zeta) + \xi(\phi-\zeta) - \gamma_{ij}u_{i,j}, \\
\varepsilon s &= (\ell_{ij}\phi_{,j})_{,i} - \xi(\phi-\zeta) - \omega_{ij}u_{i,j}, \\
a\theta &= (r_{ij}\psi_{,j})_{,i} - a_{ij}u_{i,j}.
\end{aligned} \tag{6.44}
$$

We use a logarithmic convexity argument with the function H defined by

$$
\begin{aligned}
H(t) = &< \rho u_i u_i > + \int_0^t < k_{ij}\eta_{,i}\eta_{,j} > da + \int_0^t < m_{ij}\zeta_{,i}\zeta_{,j} > da \\
&+ \int_0^t < \ell_{ij}\phi_{,i}\phi_{,j} > da + \int_0^t < r_{ij}\psi_{,i}\psi_{,j} > da \\
&+ \int_0^t < \gamma(\eta-\zeta)^2 > da + \int_0^t < \xi(\phi-\zeta)^2 > da.
\end{aligned} \tag{6.45}
$$

One differentiates H to obtain

$$
\begin{aligned}
H'(t) = &2 < \rho u_i \dot{u}_i > + 2\int_0^t < k_{ij}\eta_{,i}p_{,j} > da + 2\int_0^t < m_{ij}\zeta_{,i}q_{,j} > da \\
&+ 2\int_0^t < \ell_{ij}\phi_{,i}s_{,j} > da + 2\int_0^t < r_{ij}\psi_{,i}\theta_{,j} > da \\
&+ 2\int_0^t < \gamma(\eta-\zeta)(p-q) > da + 2\int_0^t < \xi(\phi-\zeta)(s-q) > da.
\end{aligned} \tag{6.46}
$$

Upon further differentiation one finds

$$
\begin{aligned}
H'' = &2 < \rho \dot{u}_i \dot{u}_i > + 2 < \rho u_i \ddot{u}_i > + 2 < k_{ij}\eta_{,i}p_{,j} > \\
&+ 2 < m_{ij}\zeta_{,i}q_{,j} > + 2 < \ell_{ij}\phi_{,i}s_{,j} > + 2 < r_{ij}\psi_{,i}\theta_{,j} > \\
&+ 2 < \gamma(\eta-\zeta)(p-q) > + 2 < \xi(\phi-\zeta)(s-q) > .
\end{aligned} \tag{6.47}
$$

We next multiply equation $(6.38)_1$ by u_i and integrate over Ω. We further multiply equations $(6.44)_{1-4}$ by p, q, s and θ, respectively, and integrate each over Ω. This yields the relations,

$$\begin{aligned} &< \rho \ddot{u}_i u_i > + < a_{ijkh} u_{i,j} u_{k,h} > = < \beta_{ij} p u_{i,j} > \\ &\quad + < \gamma_{ij} q u_{i,j} > + < \omega_{ij} s u_{i,j} > + < a_{ij} \theta u_{i,j} >, \end{aligned} \tag{6.48}$$

and

$$< \alpha p^2 > + < k_{ij} \eta_{,j} p_{,i} > + < \gamma p(\eta - \zeta) > + < p \beta_{ij} u_{i,j} > = 0, \tag{6.49}$$

and

$$\begin{aligned} &< \beta q^2 > + < m_{ij} \zeta_{,j} q_{,i} > - < \gamma q(\eta - \zeta) > \\ &\quad - < \xi q(\phi - \zeta) > + < q \gamma_{ij} u_{i,j} > = 0, \end{aligned} \tag{6.50}$$

and

$$< \varepsilon s^2 > + < \ell_{ij} \phi_{,j} s_{,i} > + < \xi s(\phi - \zeta) > + < s \omega_{ij} u_{i,j} > = 0, \tag{6.51}$$

and

$$< a \theta^2 > + < r_{ij} \psi_{,j} \theta_{,i} > + < a_{ij} u_{i,j} \theta > = 0. \tag{6.52}$$

Addition of equations (6.48)-(6.52) leads to the useful relation

$$\begin{aligned} &< \rho \ddot{u}_i u_i > + < k_{ij} \eta_{,j} p_{,i} > + < m_{ij} \zeta_{,j} q_{,i} > + < \ell_{ij} \phi_{,j} s_{,i} > \\ &\quad + < r_{ij} \psi_{,j} \theta_{,i} > + < \gamma(p - q)(\eta - \zeta) > + < \xi(s - q)(\phi - \zeta) > \\ &\quad = - < a_{ijkh} u_{i,j} u_{k,h} > - < \alpha p^2 > - < \beta q^2 > \\ &\quad - < \varepsilon s^2 > - < a \theta^2 > . \end{aligned} \tag{6.53}$$

We next employ (6.53) in (6.47) and thus rewrite H'' as

$$\begin{aligned} H'' =& 2 < \rho \dot{u}_i \dot{u}_i > - 2 < a_{ijkh} u_{i,j} u_{k,h} > - 2 < \alpha p^2 > \\ & - 2 < \beta q^2 > - 2 < \varepsilon s^2 > - 2 < a \theta^2 > . \end{aligned} \tag{6.54}$$

The next step involves substitution of the last five terms in (6.54) using the energy equation (6.42) and expression (6.41) to further rewrite H'' as

$$\begin{aligned} H'' =& 4 < \rho \dot{u}_i \dot{u}_i > + 4 \int_0^t < k_{ij} p_{,i} p_{,j} > da + 4 \int_0^t < m_{ij} q_{,i} q_{,j} > da \\ & + 4 \int_0^t < \ell_{ij} s_{,i} s_{,j} > da + 4 \int_0^t < r_{ij} \theta_{,i} \theta_{,j} > da \\ & + 4 \int_0^t < \gamma (p - q)^2 > da + 4 \int_0^t < \xi (s - q)^2 > da . \end{aligned} \tag{6.55}$$

Now employ (6.45), (6.46) and (6.55) to see that

$$HH'' - (H')^2 = 4S_2^2, \tag{6.56}$$

where the term S_2^2 which is non-negative by appealing to the Cauchy-Schwarz inequality is given by

$$S_2^2 = AB - C^2, \tag{6.57}$$

where the terms A, B and C are given explicitly by the following three integral expressions

$$\begin{aligned} A = &< \rho u_i u_i > + \int_0^t < k_{ij}\eta_{,i}\eta_{,j} > da + \int_0^t < m_{ij}\zeta_{,i}\zeta_{,j} > da \\ &+ \int_0^t < \ell_{ij}\phi_{,i}\phi_{,j} > da + \int_0^t < r_{ij}\psi_{,i}\psi_{,j} > da \\ &+ \int_0^t < \gamma(\eta-\zeta)^2 da > + \int_0^t < \xi(\phi-\zeta)^2 > da, \end{aligned}$$

$$\begin{aligned} B = &< \rho \dot{u}_i \dot{u}_i > + \int_0^t < k_{ij}p_{,i}p_{,j} > da + \int_0^t < m_{ij}q_{,i}q_{,j} > da \\ &+ \int_0^t < \ell_{ij}s_{,i}s_{,j} > da + \int_0^t < r_{ij}\theta_{,i}\theta_{,j} > da \\ &+ \int_0^t < \gamma(p-q)^2 > da + \int_0^t < \xi(s-q)^2 > da, \end{aligned}$$

and

$$\begin{aligned} C = &< \rho u_i \dot{u}_i > + \int_0^t < k_{ij}\eta_{,i}p_{,j} > da + \int_0^t < m_{ij}\zeta_{,i}q_{,j} > da \\ &+ \int_0^t < \ell_{ij}\phi_{,j}s_{,i} > da + \int_0^t < r_{ij}\psi_{,i}\theta_{,j} > da \\ &+ \int_0^t < \gamma(\eta-\zeta)(p-q) > da + \int_0^t < \xi(\phi-\zeta)(s-q) > da. \end{aligned}$$

From inequality (6.56) one may employ the argument of section 1.3.6 to deduce that $H \equiv 0$ on $[0,T]$, and so from (6.45), $u_i \equiv 0$ on $\Omega \times [0,T]$. Then we appeal to equations (6.42) and (6.41) to see that

$$0 \leq < \alpha p^2 > + < \beta q^2 > + < \varepsilon s^2 > + < a\theta^2 > \leq 0, \qquad t \in (0,T),$$

and so $p \equiv 0, q \equiv 0, s \equiv 0$ and $\theta \equiv 0$ on $\Omega \times [0,T]$. Thus, uniqueness of a solution to the displacement boundary - initial value problem for the equations for a linear thermoelastic body with triple porosity is proved.

It is worth drawing attention to the fact that while the three different theories discussed in this chapter are in some ways similar, the functionals used to achieve a uniqueness proof are very different. The function F in (6.11) has five terms, G in (6.28) has six, and H from (6.45) has seven terms. Even though the number of partial differential equations for the thermoelastic double porosity case and the triple porosity situation are the same, namely six equations, the functions required to establish uniqueness are different. This is because of the different number of connectivity terms between the double and triple porosity cases.

6.4 Exercises

Exercise 6.1. Consider the triple porosity thermoelasticity equations (2.46) with full connectivity between the macro, meso and micro structure porous systems, and with cross inertia terms, i.e. let u_i, p, q, s and θ satisfy the system of equations

$$\begin{aligned}
\rho\ddot{u}_i &= (a_{ijkh}u_{k,h})_{,j} - (\beta_{ij}p)_{,j} - (\gamma_{ij}q)_{,j} - (\omega_{ij}s)_{,j} - (a_{ij}\theta)_{,j} + \rho f_i\,,\\
\alpha\dot{p} + \alpha_1\dot{q} + \alpha_2\dot{s} + \alpha_3\dot{\theta} &= (k_{ij}p_{,j})_{,i} - \beta_{ij}\dot{u}_{i,j} - \gamma(p-q) - \omega(p-s) + \rho s_1\,,\\
\alpha_1\dot{p} + \beta\dot{q} + \beta_1\dot{s} + \beta_2\dot{\theta} &= (m_{ij}q_{,j})_{,i} - \gamma_{ij}\dot{u}_{i,j} + \gamma(p-q) - \xi(q-s) + \rho s_2\,,\\
\alpha_2\dot{p} + \beta_1\dot{q} + \varepsilon\dot{s} + \varepsilon_1\dot{\theta} &= (\ell_{ij}q_{,j})_{,i} - \omega_{ij}\dot{u}_{i,j} + \xi(q-s) + \omega(p-s) + \rho s_3\,,\\
\alpha_3\dot{p} + \beta_2\dot{q} + \varepsilon_1\dot{s} + a\dot{\theta} &= (r_{ij}\theta_{,j})_{,i} - a_{ij}\dot{u}_{i,j} + \rho r\,.
\end{aligned} \tag{6.58}$$

Suppose a_{ijkh} satisfy the symmetries (6.1) only, and $\beta_{ij}, \gamma_{ij}, \omega_{ij}, a_{ij}, k_{ij}, m_{ij}, \ell_{ij}$ and r_{ij} are symmetric tensors with $k_{ij}, m_{ij}, \ell_{ij}$ and r_{ij} positive. The coefficients α, β, ε, a, γ, ξ and ω may depend on $\mathbf{x}$ but are positive. Let the solution (u_i, p, q, s, θ) satisfy the boundary conditions

$$\begin{aligned}
&u_i(\mathbf{x},t) = u_i^B(\mathbf{x},t), \quad p(\mathbf{x},t) = p^B(\mathbf{x},t), \quad q(\mathbf{x},t) = q^B(\mathbf{x},t),\\
&s(\mathbf{x},t) = s^B(\mathbf{x},t), \quad \theta(\mathbf{x},t) = \theta^B(\mathbf{x},t),
\end{aligned} \tag{6.59}$$

on $\Gamma \times (0,T)$, and the initial conditions

$$\begin{aligned}
&u_i(\mathbf{x},0) = v_i(\mathbf{x}), \quad \dot{u}_i(\mathbf{x},0) = w_i(\mathbf{x}), \quad p(\mathbf{x},0) = P(\mathbf{x}),\\
&q(\mathbf{x},0) = Q(\mathbf{x}), \quad s(\mathbf{x},0) = S(\mathbf{x}), \quad \theta(\mathbf{x},0) = R(\mathbf{x}),
\end{aligned} \tag{6.60}$$

$\mathbf{x} \in \Omega$.

Show that the solution to the boundary-initial value problem $\mathscr{P}$ given by (6.58), (6.59) and (6.60) is unique.
Hint. With (u_i, p, q, s, θ) denoting the difference solution to $\mathscr{P}$ for two solutions satisfying the same source functions, the same boundary and initial conditions, use a logarithmic convexity argument with the function

$$
\begin{aligned}
F(t) = &< \rho u_i u_i > + \int_0^t < k_{ij}\eta_{,i}\eta_{,j} > da + \int_0^t < m_{ij}\zeta_{,i}\zeta_{,j} > da \\
&+ \int_0^t < \ell_{ij}\phi_{,i}\phi_{,j} > da + \int_0^t < r_{ij}\psi_{,i}\psi_{,j} > da \\
&+ \int_0^t < \gamma(\eta-\zeta)^2 > da + \int_0^t < \xi(\phi-\zeta)^2 > da \\
&+ \int_0^t < \omega(\eta-\phi)^2 > da,
\end{aligned}
$$

where η, ζ, ϕ and ψ are defined by

$$
\begin{aligned}
\eta(\mathbf{x},t) &= \int_0^t p(\mathbf{x},a)\,da, & \zeta(\mathbf{x},t) &= \int_0^t q(\mathbf{x},a)\,da, \\
\phi(\mathbf{x},t) &= \int_0^t s(\mathbf{x},a)\,da, & \psi(\mathbf{x},t) &= \int_0^t \theta(\mathbf{x},a)\,da,
\end{aligned}
$$

cf. Straughan [209].

Exercise 6.2. Consider the boundary-initial value problem, $\mathscr{P}$, for the double voids system (4.103) with a Kelvin-Voigt viscoelastic term added to the momentum equation, namely

$$
\begin{aligned}
\rho\ddot{u}_i &= (a_{ijkh}u_{k,h})_{,j} + (b_{ijkh}\dot{u}_{k,h})_{,j} + (b_{ij}\nu)_{,j} + (d_{ij}\omega)_{,j} + \rho f_i\,, \\
\kappa_1\ddot{\nu} &= (\alpha_{ij}\nu_{,j})_{,i} + (\beta_{ij}\omega_{,j})_{,i} - \alpha_1\nu - \alpha_3\omega - b_{ij}u_{i,j} + \rho\ell_1\,, \\
\kappa_2\ddot{\omega} &= (\gamma_{ij}\omega_{,j})_{,i} + (\beta_{ij}\nu_{,j})_{,i} - \alpha_2\omega - \alpha_3\nu - d_{ij}u_{i,j} + \rho\ell_2\,,
\end{aligned}
\tag{6.61}
$$

in $\Omega \times (0,T)$, with the boundary conditions

$$
\begin{aligned}
&u_i(\mathbf{x},t) = u_i^B(\mathbf{x},t), \quad \nu(\mathbf{x},t) = \nu^B(\mathbf{x},t), \\
&\omega(\mathbf{x},t) = \omega^B(\mathbf{x},t),
\end{aligned}
\tag{6.62}
$$

on $\Gamma \times (0,T)$, and the initial conditions

$$
\begin{aligned}
u_i(\mathbf{x},0) &= v_i(\mathbf{x}), & \dot{u}_i(\mathbf{x},0) &= w_i(\mathbf{x}), \\
\nu(\mathbf{x},0) &= a_1(\mathbf{x}), & \dot{\nu}(\mathbf{x},0) &= a_2(\mathbf{x}), \\
\omega(\mathbf{x},0) &= b_1(\mathbf{x}), & \dot{\omega}(\mathbf{x},0) &= b_2(\mathbf{x}),
\end{aligned}
\tag{6.63}
$$

$\mathbf{x} \in \Omega$, where $u_i^B, \nu^B, \omega^B, v_i, w_i, a_1, a_2, b_1$ and b_2 are prescribed. The functions ρ, κ_1, κ_2 are positive in Ω, the tensors $b_{ij}, d_{ij}, \alpha_{ij}, \beta_{ij}$ and γ_{ij} are symmetric, the coefficients a_{ijkh} and b_{ijkh} satisfy the symmetries (6.1) and b_{ijkh} is positive.

Show that the solution to $\mathscr{P}$ is unique.

Hint. With (u_i, ν, ω) denoting the difference solution to $\mathscr{P}$ for two solutions satisfying the same source functions, the same boundary and initial conditions, use a logarithmic convexity argument with the function

$$F(t) = < \rho u_i u_i > + < \kappa_1 v^2 > + < \kappa_2 \omega^2 > + \int_0^t < b_{ijkh} u_{i,j} u_{k,h} > ds.$$

You should show

$$F'(t) = 2 < \rho u_i \dot{u}_i > +2 < \kappa_1 v \dot{v} > +2 < \kappa_2 \omega \dot{\omega} > +2 \int_0^t < b_{ijkh} u_{i,j} \dot{u}_{k,h} > ds,$$

and

$$\begin{aligned} F'' = & 4 < \rho \dot{u}_i \dot{u}_i > +4 < \kappa_1 \dot{v}^2 > +4 < \kappa_2 \dot{\omega}^2 > -4E(0) \\ & +4 \int_0^t < b_{ijkh} \dot{u}_{i,j} \dot{u}_{k,h} > ds, \end{aligned}$$

where $E(0)$ is the energy at $t = 0$.

Exercise 6.3. In addition to double and triple porosity elasticity one may develop models to describe elasticity with a four or five porosity structure, or to describe thermoelasticity with a four or five porosity structure, cf. for example, Straughan [207]. For a four porosity structure with an isothermal elastic body we let u_i be the solid displacement, p, q, s and v be the pressures in the macro, meso, micro and sub-micro porosity systems, and then we may take as equations, see e.g. Straughan [207], equations (7),

$$\begin{aligned} \rho \ddot{u}_i = & (a_{ijkh} u_{k,h})_{,j} - (\beta_{ij} p)_{,j} - (\gamma_{ij} q)_{,j} \\ & - (\omega_{ij} s)_{,j} - (\alpha_{ij} v)_{,j} + \rho f_i, \\ \alpha \dot{p} = & (k_{ij} p_{,j})_{,i} - \beta_{ij} \dot{u}_{i,j} - \gamma(p - q) + s_1, \\ \beta \dot{q} = & (m_{ij} q_{,j})_{,i} - \gamma_{ij} \dot{u}_{i,j} + \gamma(p - q) - \xi(q - s) + s_2, \\ \varepsilon \dot{s} = & (\ell_{ij} s_{,j})_{,i} - \omega_{ij} \dot{u}_{i,j} + \xi(q - s) - \hat{\delta}(s - v) + s_3, \\ \phi \dot{v} = & (q_{ij} v_{,j})_{,i} - \alpha_{ij} \dot{u}_{i,j} + \hat{\delta}(s - v) + s_4. \end{aligned} \tag{6.64}$$

Consider the boundary-initial value problem $\mathscr{P}$ consisting of (6.64) together with the boundary conditions

$$\begin{aligned} & u_i(\mathbf{x},t) = u_i^B(\mathbf{x},t), \quad p(\mathbf{x},t) = p^B(\mathbf{x},t), \quad q(\mathbf{x},t) = q^B(\mathbf{x},t), \\ & s(\mathbf{x},t) = s^B(\mathbf{x},t), \quad v(\mathbf{x},t) = v^B(\mathbf{x},t), \qquad \text{on } \Gamma \times (0,T), \end{aligned} \tag{6.65}$$

and the initial conditions

$$\begin{aligned} & u_i(\mathbf{x},0) = v_i(\mathbf{x}), \quad \dot{u}_i(\mathbf{x},0) = w_i(\mathbf{x}), \quad p(\mathbf{x},0) = P(\mathbf{x}), \\ & q(\mathbf{x},0) = Q(\mathbf{x}), \quad s(\mathbf{x},0) = S(\mathbf{x}), \quad v(\mathbf{x},0) = V(\mathbf{x}), \qquad \mathbf{x} \in \Omega, \end{aligned} \tag{6.66}$$

where the functions $f_i, s_1, \ldots, s_4, u_i^B, p^B, q^B, s^B, v^B, v_i, w_i, P, Q, S$ and V are prescribed.

Use the ideas of this chapter to construct a suitable function and employ a logarithmic convexity method to show the solution to $\mathscr{P}$ is unique.

Exercise 6.4. For a four porosity structure in a thermoelastic body we let u_i be the solid displacement, p,q,s and v be the pressures in the macro, meso, micro and sub-micro porosity systems, and we let θ be the temperature in the body. A suitable system of equations to describe this scenario may be taken to be

$$\begin{aligned}
\rho\ddot{u}_i =&(a_{ijkh}u_{k,h})_{,j} - (\beta_{ij}p)_{,j} - (\gamma_{ij}q)_{,j}\\
&- (\omega_{ij}s)_{,j} - (\alpha_{ij}v)_{,j} - (a_{ij}\theta)_{,j} + \rho f_i\,,\\
\alpha\dot{p} =&(k_{ij}p_{,j})_{,i} - \beta_{ij}\dot{u}_{i,j} - \gamma(p-q) + s_1\,,\\
\beta\dot{q} =&(m_{ij}q_{,j})_{,i} - \gamma_{ij}\dot{u}_{i,j} + \gamma(p-q) - \xi(q-s) + s_2\,,\\
\varepsilon\dot{s} =&(\ell_{ij}s_{,j})_{,i} - \omega_{ij}\dot{u}_{i,j} + \xi(q-s) - \hat{\delta}(s-v) + s_3\,,\\
\phi\dot{v} =&(q_{ij}v_{,j})_{,i} - \alpha_{ij}\dot{u}_{i,j} + \hat{\delta}(s-v) + s_4\,,\\
a\dot{\theta} =&(r_{ij}\theta_{,j})_{,i} - a_{ij}\dot{u}_{i,j} + \rho r.
\end{aligned} \tag{6.67}$$

Consider the boundary-initial value problem $\mathscr{P}$ consisting of (6.67) together with the boundary conditions

$$\begin{aligned}
&u_i(\mathbf{x},t) = u_i^B(\mathbf{x},t), \quad p(\mathbf{x},t) = p^B(\mathbf{x},t),\\
&q(\mathbf{x},t) = q^B(\mathbf{x},t), \quad s(\mathbf{x},t) = s^B(\mathbf{x},t),\\
&v(\mathbf{x},t) = v^B(\mathbf{x},t), \quad \theta(\mathbf{x},t) = \theta^B(\mathbf{x},t), \qquad \text{on } \Gamma\times(0,T),
\end{aligned} \tag{6.68}$$

and the initial conditions

$$\begin{aligned}
&u_i(\mathbf{x},0) = v_i(\mathbf{x}), \quad \dot{u}_i(\mathbf{x},0) = w_i(\mathbf{x}), \quad p(\mathbf{x},0) = P(\mathbf{x}), \quad q(\mathbf{x},0) = Q(\mathbf{x}),\\
&s(\mathbf{x},0) = S(\mathbf{x}), \quad v(\mathbf{x},0) = V(\mathbf{x}), \quad \theta(\mathbf{x},0) = R(\mathbf{x}), \qquad \mathbf{x}\in\Omega,
\end{aligned} \tag{6.69}$$

where the functions $f_i, s_1, \ldots, s_4, u_i^B, p^B, q^B, s^B, v^B, \theta^B, v_i, w_i, P, Q, S, V$ and R are prescribed.

Use the ideas of this chapter to construct a suitable function and employ a logarithmic convexity method to show the solution to $\mathscr{P}$ is unique.

Exercise 6.5. When the material has a five porosity structure with an isothermal elastic body we let u_i be the solid displacement, p,q,s,v and w be the pressures in the macro, meso, micro, sub-micro and nano porosity systems, and then we may take as equations, see e.g. Straughan [207], equations (10),

$$\begin{aligned}
\rho\ddot{u}_i =&(a_{ijkh}u_{k,h})_{,j} - (\beta_{ij}p)_{,j} - (\gamma_{ij}q)_{,j}\\
&- (\omega_{ij}s)_{,j} - (\alpha_{ij}v)_{,j} - (\eta_{ij}w)_{,j} + \rho f_i\,,\\
\alpha\dot{p} =&(k_{ij}p_{,j})_{,i} - \beta_{ij}\dot{u}_{i,j} - \gamma(p-q) + s_1\,,\\
\beta\dot{q} =&(m_{ij}q_{,j})_{,i} - \gamma_{ij}\dot{u}_{i,j} + \gamma(p-q) - \xi(q-s) + s_2\,,\\
\varepsilon\dot{s} =&(\ell_{ij}s_{,j})_{,i} - \omega_{ij}\dot{u}_{i,j} + \xi(q-s) - \hat{\delta}(s-v) + s_3\,,\\
\phi\dot{v} =&(q_{ij}v_{,j})_{,i} - \alpha_{ij}\dot{u}_{i,j} + \hat{\delta}(s-v) - \zeta(v-w) + s_4\,,\\
\eta\dot{w} =&(p_{ij}w_{,j})_{,i} - \eta_{ij}\dot{u}_{i,j} + \zeta(v-w) + s_5\,.
\end{aligned} \tag{6.70}$$

Consider the boundary-initial value problem $\mathscr{P}$ consisting of (6.70) together with the boundary conditions

$$\begin{aligned} &u_i(\mathbf{x},t)=u_i^B(\mathbf{x},t), \quad p(\mathbf{x},t)=p^B(\mathbf{x},t), \quad q(\mathbf{x},t)=q^B(\mathbf{x},t),\\ &s(\mathbf{x},t)=s^B(\mathbf{x},t), \quad v(\mathbf{x},t)=v^B(\mathbf{x},t), \quad w(\mathbf{x},t)=w^B(\mathbf{x},t), \end{aligned} \tag{6.71}$$

on $\Gamma\times(0,T)$, and the initial conditions

$$\begin{aligned} &u_i(\mathbf{x},0)=v_i(\mathbf{x}), \quad \dot{u}_i(\mathbf{x},0)=w_i(\mathbf{x}), \quad p(\mathbf{x},0)=P(\mathbf{x}),\\ &q(\mathbf{x},0)=Q(\mathbf{x}), \quad s(\mathbf{x},0)=S(\mathbf{x}), \quad v(\mathbf{x},0)=V(\mathbf{x}), \quad w(\mathbf{x},0)=W(\mathbf{x}), \end{aligned} \tag{6.72}$$

for $\mathbf{x}\in\Omega$, where the functions $f_i,s_1,\ldots,s_5,u_i^B,p^B,q^B,s^B,v^B,w^B,v_i,w_i,P,Q,S,V$ and W are prescribed.

Use the ideas of this chapter to construct a suitable function and employ a logarithmic convexity method to show the solution to $\mathscr{P}$ is unique.

Exercise 6.6. For a five porosity structure in a thermoelastic body we let u_i be the solid displacement, p,q,s,v and w be the pressures in the macro, meso, micro, submicro and nano porosity systems, and we let θ be the temperature in the body. A suitable system of equations to describe this scenario may be taken to be

$$\begin{aligned} \rho\ddot{u}_i =&(a_{ijkh}u_{k,h})_{,j}-(\beta_{ij}p)_{,j}-(\gamma_{ij}q)_{,j}\\ &-(\omega_{ij}s)_{,j}-(\alpha_{ij}v)_{,j}-(\eta_{ij}w)_{,j}-(a_{ij}\theta)_{,j}+\rho f_i\,,\\ \alpha\dot{p} =&(k_{ij}p_{,j})_{,i}-\beta_{ij}\dot{u}_{i,j}-\gamma(p-q)+s_1\,,\\ \beta\dot{q} =&(m_{ij}q_{,j})_{,i}-\gamma_{ij}\dot{u}_{i,j}+\gamma(p-q)-\xi(q-s)+s_2\,,\\ \varepsilon\dot{s} =&(\ell_{ij}s_{,j})_{,i}-\omega_{ij}\dot{u}_{i,j}+\xi(q-s)-\hat{\delta}(s-v)+s_3\,,\\ \phi\dot{v} =&(q_{ij}v_{,j})_{,i}-\alpha_{ij}\dot{u}_{i,j}+\hat{\delta}(s-v)-\zeta(v-w)+s_4\,,\\ \eta\dot{w} =&(p_{ij}w_{,j})_{,i}-\eta_{ij}\dot{u}_{i,j}+\zeta(v-w)+s_5\,,\\ a\dot{\theta} =&(r_{ij}\theta_{,j})_{,i}-a_{ij}\dot{u}_{i,j}+\rho r. \end{aligned} \tag{6.73}$$

Consider the boundary-initial value problem $\mathscr{P}$ consisting of (6.73) together with the boundary conditions

$$\begin{aligned} &u_i(\mathbf{x},t)=u_i^B(\mathbf{x},t), \quad p(\mathbf{x},t)=p^B(\mathbf{x},t), \quad q(\mathbf{x},t)=q^B(\mathbf{x},t),\\ &s(\mathbf{x},t)=s^B(\mathbf{x},t), \quad v(\mathbf{x},t)=v^B(\mathbf{x},t),\\ &w(\mathbf{x},t)=w^B(\mathbf{x},t), \quad \theta(\mathbf{x},t)=\theta^B(\mathbf{x},t), \end{aligned} \tag{6.74}$$

on $\Gamma\times(0,T)$, and the initial conditions

$$\begin{aligned} &u_i(\mathbf{x},0)=v_i(\mathbf{x}), \quad \dot{u}_i(\mathbf{x},0)=w_i(\mathbf{x}), \quad p(\mathbf{x},0)=P(\mathbf{x}), \quad q(\mathbf{x},0)=Q(\mathbf{x}),\\ &s(\mathbf{x},0)=S(\mathbf{x}), \quad v(\mathbf{x},0)=V(\mathbf{x}), \quad w(\mathbf{x},0)=W(\mathbf{x}), \quad \theta(\mathbf{x},0)=R(\mathbf{x}), \end{aligned} \tag{6.75}$$

for $\mathbf{x} \in \Omega$, where the functions $f_i, s_1, \ldots, s_5, u_i^B, p^B, q^B, s^B, v^B, w^B, \theta^B$, v_i, w_i, P, Q, S, V, W and R are prescribed.

Use the ideas of this chapter to construct a suitable function and employ a logarithmic convexity method to show the solution to $\mathscr{P}$ is unique.

Exercise 6.7. Quintanilla [175] deals with a system of equations for an isotropic thermoelastic body saturated with either a gas or a fluid. In the case of fluid saturation the system of equations may be written for an *anisotropic* thermoelastic body as

$$\begin{aligned} &\rho_w \ddot{w}_i = a_4 w_{j,ji} + a_5 u_{j,ji} + \beta_2 T_{,i} - \xi_3(\dot{w}_i - \dot{u}_i) + (b_{ijkh}\dot{w}_{k,h})_{,j}\,, \\ &\rho_u \ddot{u}_i = a_5 w_{j,ji} + (a_{ijkh}u_{k,h})_{,j} + \beta_3 T_{,i} + \xi_3(\dot{w}_i - \dot{u}_i)\,, \\ &c\dot{T} = \beta_2 \dot{w}_{j,j} + \beta_3 \dot{u}_{j,j} + (k_{ij}T_{,j})_{,i}\,, \end{aligned} \tag{6.76}$$

in $\Omega \times (0,T)$, where w_i, u_i and T denote the fluid displacement, solid displacement, and temperature, respectively. Supoose a_{ijkh} and b_{ijkh} satisfy the symmetries (6.1) with b_{ijkh} also being positive. The tensor k_{ij} is symmetric and positive. For simplicity we take the coefficients $\rho_w, \rho_u, c, a_4, a_5, \beta_2$ and β_3 to be constants while $\xi_3(\mathbf{x})$ is positive for all $\mathbf{x} \in \Omega$. Let the solution to (6.76) satisfy the boundary conditions

$$\begin{aligned} &u_i(\mathbf{x},t) = u_i^B(\mathbf{x},t), \quad w_i(\mathbf{x},t) = w_i^B(\mathbf{x},t), \\ &T(\mathbf{x},t) = T^B(\mathbf{x},t), \qquad \text{on } \Gamma \times (0,T), \end{aligned} \tag{6.77}$$

for some $T < \infty$, and the initial conditions

$$\begin{aligned} &u_i(\mathbf{x},0) = \psi_i^1(\mathbf{x}), \quad \dot{u}_i(\mathbf{x},0) = \psi_i^2(\mathbf{x}), \\ &w_i(\mathbf{x},0) = \phi_i^1(\mathbf{x}), \quad \dot{w}_i(\mathbf{x},0) = \phi_i^2(\mathbf{x}), \\ &T(\mathbf{x},0) = T_0(\mathbf{x}) \qquad \mathbf{x} \in \Omega\,. \end{aligned} \tag{6.78}$$

Show that the solution to the boundary-initial value problem $\mathscr{P}$ given by (6.76), (6.77) and (6.78) is unique.
Hint. Use a logarithmic convexity argument with the function

$$\begin{aligned} H(t) = &< \rho_u u_i u_i > + < \rho_w w_i w_i > + \int_0^t < k_{ij}\eta_{,i}\eta_{,j} > ds \\ &+ \int_0^t < b_{ijkh} w_{i,j} w_{k,h} > ds + \int_0^t < \xi_3(w_i - u_i)(w_i - u_i) > ds, \end{aligned} \tag{6.79}$$

where $\eta(\mathbf{x},t) = \int_0^t T(\mathbf{x},s)ds$, where in (6.79) the functions u_i, w_i and η refer to the difference solution to two solutions to $\mathscr{P}$.

You should find

$$\begin{aligned} H' = &2 < \rho_u u_i \dot{u}_i > + 2 < \rho_w w_i \dot{w}_i > + 2\int_0^t < k_{ij}\eta_{,i}T_{,j} > ds \\ &+ 2\int_0^t < b_{ijkh} w_{i,j} \dot{w}_{k,h} > ds + 2\int_0^t < \xi_3(w_i - u_i)(\dot{w}_i - \dot{u}_i) > ds, \end{aligned}$$

and after using the energy equation and the time integrated form of equation $(6.76)_3$ you should derive

$$\begin{aligned}H'' =& 4 < \rho_u \dot{u}_i \dot{u}_i > +4 < \rho_w \dot{w}_i \dot{w}_i > +4 \int_0^t < b_{ijkh} \dot{w}_{i,j} \dot{w}_{k,h} > ds \\ &+4 \int_0^t < \xi_3 (\dot{w}_i - \dot{u}_i)(\dot{w}_i - \dot{u}_i) > ds + 4 \int_0^t < k_{ij} T_{,i} T_{,j} > ds - 4E(0),\end{aligned}$$

where $E(0)$ is the energy at $t = 0$.

Exercise 6.8. Quintanilla [175] deals with a system of equations for an isotropic thermoelastic body saturated with either a gas or a fluid. In the case of gas saturation the system of equations may be written for an *anisotropic* thermoelastic body as

$$\begin{aligned}&\rho_z \ddot{z}_i = a_1 z_{j,ji} + a_3 u_{j,ji} + \beta_1 T_{,i} - \xi_1 (\dot{z}_i - \dot{u}_i), \\ &\rho_u \ddot{u}_i = a_3 z_{j,ji} + (a_{ijkh} u_{k,h})_{,j} + \beta_3 T_{,i} + \xi_1 (\dot{z}_i - \dot{u}_i), \\ &c \dot{T} = \beta_1 \dot{z}_{j,j} + \beta_3 \dot{u}_{j,j} + (k_{ij} T_{,j})_{,i},\end{aligned} \tag{6.80}$$

in $\Omega \times (0,T)$, where z_i, u_i and T denote the gas particle displacement, solid displacement, and temperature, respectively. Supoose a_{ijkh} satisfy the symmetries (6.1). The tensor k_{ij} is symmetric and positive. For simplicity we take the coefficients $\rho_z, \rho_u, c, a_1, a_3, \beta_1$ and β_3 to be constants while $\xi_1(\mathbf{x})$ is positive for all $\mathbf{x} \in \Omega$. Let the solution to (6.80) satisfy the boundary conditions

$$\begin{aligned}&u_i(\mathbf{x},t) = u_i^B(\mathbf{x},t), \quad z_i(\mathbf{x},t) = z_i^B(\mathbf{x},t), \\ &T(\mathbf{x},t) = T^B(\mathbf{x},t), \qquad \text{on } \Gamma \times (0,T),\end{aligned} \tag{6.81}$$

for some $T < \infty$, and the initial conditions

$$\begin{aligned}&u_i(\mathbf{x},0) = \psi_i^1(\mathbf{x}), \quad \dot{u}_i(\mathbf{x},0) = \psi_i^2(\mathbf{x}), \\ &z_i(\mathbf{x},0) = \phi_i^1(\mathbf{x}), \quad \dot{z}_i(\mathbf{x},0) = \phi_i^2(\mathbf{x}), \\ &T(\mathbf{x},0) = T_0(\mathbf{x}), \qquad \mathbf{x} \in \Omega.\end{aligned} \tag{6.82}$$

Show that the solution to the boundary-initial value problem $\mathscr{P}$ given by (6.80), (6.81) and (6.82) is unique.
Hint. Use a logarithmic convexity argument with the function

$$\begin{aligned}F(t) =& < \rho_u u_i u_i > + < \rho_z z_i z_i > + \int_0^t < k_{ij} \eta_{,i} \eta_{,j} > ds \\ &+ \int_0^t < \xi_1 (z_i - u_i)(z_i - u_i) > ds,\end{aligned} \tag{6.83}$$

where $\eta(\mathbf{x},t) = \int_0^t T(\mathbf{x},s) ds$, where in (6.83) the functions u_i, z_i and η refer to the difference solution to two solutions to $\mathscr{P}$.

Exercise 6.9. Sharma [193] deals with a system of equations for an isotropic elastic body which is saturated with a combination of a gas and a fluid. We here write the system of equations for an *anisotropic* elastic body as

$$\begin{aligned}
\rho_1 \ddot{u}_i &= (a_{ijkh}u_{k,h})_{,j} + (a_{12}v_{j,j} + a_{13}w_{j,j})_{,i} + q_g(\dot{v}_i - \dot{u}_i) + q_\ell(\dot{w}_i - \dot{u}_i)\,, \\
\rho_2 \ddot{v}_i &= (a_{21}u_{j,j} + a_{22}v_{j,j} + a_{23}w_{j,j})_{,i} - q_g(\dot{v}_i - \dot{u}_i)\,, \\
\rho_3 \ddot{w}_i &= (a_{31}u_{j,j} + a_{32}v_{j,j} + a_{33}w_{j,j})_{,i} - q_\ell(\dot{w}_i - \dot{u}_i)\,,
\end{aligned} \tag{6.84}$$

in $\Omega \times (0,T)$, where u_i, v_i and w_i denote the solid displacement, gas particle displacement, and fluid particle displacement, respectively. Suppose a_{ijkh} satisfy the symmetries (6.1). The coefficients $\rho_1, \rho_2, \rho_3, a_{12}, \dots, a_{33}, q_g, q_\ell$ are functions of $\mathbf{x}$, and a_{22}, a_{33} are positive in Ω. Let the solution to (6.84) satisfy the boundary conditions

$$\begin{aligned}
&u_i(\mathbf{x},t) = u_i^B(\mathbf{x},t), \quad v_i(\mathbf{x},t) = v_i^B(\mathbf{x},t), \\
&w_i(\mathbf{x},t) = w_i^B(\mathbf{x},t), \qquad \text{on } \Gamma \times (0,T),
\end{aligned} \tag{6.85}$$

for some $T < \infty$, and the initial conditions

$$\begin{aligned}
u_i(\mathbf{x},0) &= \alpha_i^1(\mathbf{x}), \quad \dot{u}_i(\mathbf{x},0) = \alpha_i^2(\mathbf{x}), \\
v_i(\mathbf{x},0) &= \beta_i^1(\mathbf{x}), \quad \dot{v}_i(\mathbf{x},0) = \beta_i^2(\mathbf{x}), \\
w_i(\mathbf{x},0) &= \gamma_i^1(\mathbf{x}), \quad \dot{w}_i(\mathbf{x},0) = \gamma_i^2(\mathbf{x}), \qquad \mathbf{x} \in \Omega.
\end{aligned} \tag{6.86}$$

Show that the solution to the boundary-initial value problem $\mathscr{P}$ given by (6.84), (6.85) and (6.86) is unique.
Hint. Use a logarithmic convexity argument with the function

$$\begin{aligned}
J(t) &= < \rho_1 u_i u_i > + < \rho_2 v_i v_i > + < \rho_3 w_i w_i > \\
&\quad + \int_0^t < q_\ell (u_i - w_i)(u_i - w_i) > ds + \int_0^t < q_g (u_i - v_i)(u_i - v_i) > ds,
\end{aligned} \tag{6.87}$$

where in (6.87) the functions u_i, v_i and w_i refer to the difference solution to two solutions to $\mathscr{P}$.

Chapter 7
Continuous Dependence in Multi - Porosity Elasticity

7.1 Continuous Dependence upon the Initial Data

We commence this chapter with an analysis of continuous dependence upon the initial data for the isothermal triple porosity system (2.44) and we require only the symmetry conditions (6.1) for the elastic coefficients a_{ijkh}. Thus, we let $u_i(\mathbf{x},t)$ denote the elastic displacement, and we let $p(\mathbf{x},t), q(\mathbf{x},t)$ and $s(\mathbf{x},t)$ be the pressures in the macro, meso and micro porosity structures. The functions (u_i, p, q, s) satisfy the boundary - initial value problem

$$\begin{aligned}
\rho\ddot{u}_i &= (a_{ijkh}u_{k,h})_{,j} - (\beta_{ij}p)_{,j} - (\gamma_{ij}q)_{,j} - (\omega_{ij}s)_{,j}\,,\\
\alpha\dot{p} &= (k_{ij}p_{,j})_{,i} - \gamma(p-q) - \beta_{ij}\dot{u}_{i,j}\,,\\
\beta\dot{q} &= (m_{ij}q_{,j})_{,i} + \gamma(p-q) + \xi(s-q) - \gamma_{ij}\dot{u}_{i,j}\,,\\
\varepsilon\dot{s} &= (\ell_{ij}s_{,j})_{,i} - \xi(s-q) - \omega_{ij}\dot{u}_{i,j}\,,
\end{aligned} \tag{7.1}$$

in $\Omega \times (0,T)$, for some $0 < T < \infty$, where the boundary conditions are

$$\begin{aligned}
u_i(\mathbf{x},t) &= u_i^B(\mathbf{x},t), \qquad & p(\mathbf{x},t) &= p^B(\mathbf{x},t),\\
q(\mathbf{x},t) &= q^B(\mathbf{x},t), \qquad & s(\mathbf{x},t) &= s^B(\mathbf{x},t),
\end{aligned} \tag{7.2}$$

on $\Gamma \times (0,T)$, with the initial conditions being

$$\begin{aligned}
u_i(\mathbf{x},0) &= v_i(\mathbf{x}), \qquad & \dot{u}_i(\mathbf{x},0) &= w_i(\mathbf{x}), \qquad & p(\mathbf{x},0) &= P(\mathbf{x}),\\
q(\mathbf{x},0) &= Q(\mathbf{x}), \qquad & s(\mathbf{x},0) &= S(\mathbf{x}), \qquad & \mathbf{x} &\in \Omega.
\end{aligned} \tag{7.3}$$

The functions $u_i^B, p^B, q^B, s^B, v_i, w_i, P, Q$ and S are given. The coefficients $\rho, \alpha, \beta, \varepsilon, \gamma$ and ξ are positive, the tensors $k_{ij}, m_{ij}, \ell_{ij}, \beta_{ij}, \gamma_{ij}$, and ω_{ij} are symmetric with k_{ij}, m_{ij} and ℓ_{ij} being positive. The elastic coefficients are *not* required to be positive and we request only that they satisfy the symmetries

$$a_{ijkh} = a_{jikh} = a_{khij}\,. \tag{7.4}$$

B. Straughan, *Mathematical Aspects of Multi–Porosity Continua*, Advances in Mechanics and Mathematics 38, https://doi.org/10.1007/978-3-319-70172-1_7

We are unable to follow the analysis of section 6.2 for uniqueness because the initial data are no longer zero. The non-zero initial data require one to modify the function (6.28) employed in the uniqueness proof.

To establish continuous dependence upon the initial data we let (u_i^1, p^1, q^1, s^1) and (u_i^2, p^2, q^2, s^2) be solutions to (7.1) - (7.3) for the same boundary data but for different initial data functions $(v_i^1, w_i^1, P^1, Q^1, S^1)$ and $(v_i^2, w_i^2, P^2, Q^2, S^2)$. Define the difference functions

$$
\begin{aligned}
&u_i = u_i^1 - u_i^2, \quad p = p^1 - p^2, \quad q = q^1 - q^2, \quad s = s^1 - s^2, \quad v_i = v_i^1 - v_i^2, \\
&w_i = w_i^1 - w_i^2, \quad P = P^1 - P^2, \quad Q = Q^1 - Q^2, \quad S = S^1 - S^2,
\end{aligned}
\tag{7.5}
$$

and then one finds that (u_i, p, q, s) satisfies the boundary-initial value problem $\mathscr{P}$ given by equations (7.1) together with the boundary conditions

$$
\begin{aligned}
&u_i(\mathbf{x},t) = 0, \qquad p(\mathbf{x},t) = 0, \\
&q(\mathbf{x},t) = 0, \qquad s(\mathbf{x},t) = 0, \qquad \text{on } \Gamma \times (0,T),
\end{aligned}
\tag{7.6}
$$

while the initial conditions remain as (7.2) where now v_i, w_i, P, Q and S are defined by (7.5).

The analysis proceeds via a logarithmic convexity argument and we begin by introducing the functions η, ζ and ϕ by

$$
\eta(\mathbf{x},t) = \int_0^t p(\mathbf{x},a)da, \quad \zeta(\mathbf{x},t) = \int_0^t q(\mathbf{x},a)da, \quad \phi(\mathbf{x},t) = \int_0^t s(\mathbf{x},a)da.
$$

Then we define the functions $D(\mathbf{x}), E(\mathbf{x}), F(\mathbf{x})$ to be the solution to the system of equations

$$
\begin{aligned}
&(k_{ij}D_{,j})_{,i} - \gamma(D - E) = \alpha P(\mathbf{x}) + \beta_{ij} v_{i,j}(\mathbf{x}), \\
&(m_{ij}E_{,j})_{,i} + \gamma(D - E) + \xi(F - E) = \beta Q(\mathbf{x}) + \gamma_{ij} v_{i,j}(\mathbf{x}), \\
&(\ell_{ij}F_{,j})_{,i} - \xi(F - E) = \varepsilon S(\mathbf{x}) + \omega_{ij} v_{i,j}(\mathbf{x}).
\end{aligned}
\tag{7.7}
$$

Due to non-zero initial data it is not convenient to work directly with the functions η, ζ and ϕ. Thus, the next step is to introduce the functions $A(\mathbf{x},t), B(\mathbf{x},t)$ and $C(\mathbf{x},t)$ by

$$
A = \eta + D, \qquad B = \zeta + E, \qquad C = \phi + F. \tag{7.8}
$$

To apply a logarithmic convexity method we introduce now the function $G(t)$, which involves η, ζ and ϕ through the functions A, B and C, and G is given by

$$
\begin{aligned}
G(t) = &< \rho u_i u_i > + \int_0^t < k_{ij} A_{,i} A_{,j} > ds + \int_0^t < m_{ij} B_{,i} B_{,j} > ds \\
&+ \int_0^t < \ell_{ij} C_{,i} C_{,j} > ds + \int_0^t < \gamma (A-B)^2 > ds \\
&+ \int_0^t < \xi (C-B)^2 > ds + (T-t)\big[< k_{ij} D_{,i} D_{,j} > \\
&\quad + < m_{ij} E_{,i} E_{,j} > + < \ell_{ij} F_{,i} F_{,j} > \\
&\quad + < \gamma (D-E)^2 > + < \xi (F-E)^2 > \big].
\end{aligned} \tag{7.9}
$$

After differentiation one may show that

$$
\begin{aligned}
G' = &2 < \rho u_i \dot{u}_i > + 2 \int_0^t < k_{ij} A_{,i} p_{,j} > da + 2 \int_0^t < m_{ij} B_{,i} q_{,j} > da \\
&+ 2 \int_0^t < \ell_{ij} C_{,i} s_{,j} > da + 2 \int_0^t < \gamma (A-B)(p-q) > da \\
&+ 2 \int_0^t < \xi (C-B)(s-q) > da.
\end{aligned} \tag{7.10}
$$

One differentiates G' and then after substitution for $\rho \ddot{u}_i$ from $(7.1)_1$ and some integration by parts we find

$$
\begin{aligned}
G'' = &2 < \rho \dot{u}_i \dot{u}_i > - 2 < a_{ijkh} u_{i,j} u_{k,h} > + 2 < \beta_{ij} p u_{i,j} > \\
&+ 2 < \gamma_{ij} q u_{i,j} > + 2 < \omega_{ij} s u_{i,j} > + 2 < k_{ij} A_{,i} p_{,j} > \\
&+ 2 < m_{ij} B_{,i} q_{,j} > + 2 < \ell_{ij} C_{,i} s_{,j} > \\
&+ 2 < \gamma (A-B)(p-q) > + 2 < \xi (C-B)(s-q) > .
\end{aligned} \tag{7.11}
$$

To proceed from this point we integrate each of equations (7.1) from 0 to t to obtain

$$
\begin{aligned}
\alpha \dot{A} = \alpha p &= (k_{ij} A_{,j})_{,i} - \gamma (A-B) - \beta_{ij} u_{i,j}, \\
\beta \dot{B} = \beta q &= (m_{ij} B_{,j})_{,i} + \gamma (A-B) + \xi (C-B) - \gamma_{ij} u_{i,j}, \\
\varepsilon \dot{C} = \varepsilon s &= (\ell_{ij} C_{,j})_{,i} - \xi (C-B) - \omega_{ij} u_{i,j}.
\end{aligned} \tag{7.12}
$$

We next multiply $(7.12)_{1-3}$ by p, q, s in turn and integrate over Ω. We form the sum of the resulting three equations and after some integration by parts we may obtain

$$
\begin{aligned}
2 < \alpha p^2 > + 2 < \beta q^2 > + 2 < \varepsilon s^2 > = &- 2 < k_{ij} A_{,i} p_{,j} > \\
&- 2 < m_{ij} B_{,j} q_{,i} > - 2 < \ell_{ij} C_{,j} s_{,i} > - 2 < \beta_{ij} u_{i,j} p > \\
&- 2 < \gamma_{ij} u_{i,j} q > - 2 < \omega_{ij} u_{i,j} s > \\
&- 2 < \gamma (A-B)(p-q) > - 2 < \xi (C-B)(s-q) > .
\end{aligned} \tag{7.13}
$$

The next step is to employ (7.13) in (7.11) to arrive at

$$
\begin{aligned}
G'' = &2 < \rho \dot{u}_i \dot{u}_i > - 2 < a_{ijkh} u_{i,j} u_{k,h} > - 2 < \alpha p^2 > \\
&- 2 < \beta q^2 > - 2 < \varepsilon s^2 > .
\end{aligned} \tag{7.14}
$$

We now need the energy equation and from (7.1) this may be shown to be

$$\begin{aligned} E(t)+&\int_0^t < k_{ij}p_{,i}p_{,j} > da+\int_0^t < m_{ij}q_{,i}q_{,j} > da\\ +&\int_0^t < \ell_{ij}s_{,i}s_{,j} > da+\int_0^t < \gamma(p-q)^2 > da\\ +&\int_0^t < \xi(s-q)^2 > da=E(0)=0, \end{aligned} \tag{7.15}$$

where

$$\begin{aligned} E(t)=&\frac{1}{2} < \rho\dot{u}_i\dot{u}_i > +\frac{1}{2} < a_{ijkh}u_{i,j}u_{k,h} > +\frac{1}{2} < \alpha p^2 >\\ &+\frac{1}{2} < \beta q^2 > +\frac{1}{2} < \varepsilon s^2 > . \end{aligned} \tag{7.16}$$

Finally we substitute for the last four terms in (7.14) using (7.15) and (7.16) to arrive at

$$\begin{aligned} G''=&4 < \rho\dot{u}_i\dot{u}_i > +4\int_0^t < k_{ij}p_{,i}p_{,j} > da+4\int_0^t < m_{ij}q_{,i}q_{,j} > da\\ &+4\int_0^t < \ell_{ij}s_{,i}s_{,j} > da+4\int_0^t < \gamma(p-q)^2 > da\\ &+4\int_0^t < \xi(s-q)^2 > da-4E(0). \end{aligned} \tag{7.17}$$

We are now in a position to form $GG''-(G')^2$ using (7.9), (7.10) and (7.17). This leads to

$$\begin{aligned} GG''-(G')^2=&4S^2-4E(0)G+4(T-t)\big[< k_{ij}D_{,i}D_{,j} > + < m_{ij}E_{,i}E_{,j} >\\ &+ < \ell_{ij}F_{,i}F_{,j} > + < \gamma(D-E)^2 > + < \xi(F-E)^2 >\big]\\ &\times\bigg[< \rho\dot{u}_i\dot{u}_i > +\int_0^t < k_{ij}p_{,i}p_{,j} > da+\int_0^t < m_{ij}q_{,i}q_{,j} > da\\ &+\int_0^t < \ell_{ij}s_{,i}s_{,j} > da+\int_0^t < \gamma(p-q)^2 > da\\ &+\int_0^t < \xi(s-q)^2 > da\bigg], \end{aligned} \tag{7.18}$$

where

$$S^2=XY-Z^2, \tag{7.19}$$

and where the terms X,Y and Z are given by the following expressions

$$X = < \rho u_i u_i > + \int_0^t < k_{ij} A_{,i} A_{,j} > da$$
$$+ \int_0^t < m_{ij} B_{,i} B_{,j} > da + \int_0^t < \ell_{ij} C_{,i} C_{,j} > da$$
$$+ \int_0^t < \gamma (A-B)^2 > da + \int_0^t < \xi (C-B)^2 > da$$

and

$$Y = < \rho \dot{u}_i \dot{u}_i > + \int_0^t < k_{ij} p_{,i} p_{,j} > da$$
$$+ \int_0^t < m_{ij} q_{,i} q_{,j} > da + \int_0^t < \ell_{ij} s_{,i} s_{,j} > ds$$
$$+ \int_0^t < \gamma (p-q)^2 > da + \int_0^t < \xi (s-q)^2 > da$$

and

$$Z = < \rho u_i \dot{u}_i > + \int_0^t < k_{ij} A_{,i} p_{,j} > da$$
$$+ \int_0^t < m_{ij} B_{,i} q_{,j} > da + \int_0^t < \ell_{ij} C_{,i} s_{,j} > da$$
$$+ \int_0^t < \gamma (A-B)(p-q) > da + \int_0^t < \xi (C-B)(s-q) > da .$$

From (7.18) we deduce that

$$GG'' - (G')^2 \geq -4E(0)G. \tag{7.20}$$

7.1.1 Case of $E(0) \leq 0$

When $E(0) \leq 0$ then from (7.20) we find $(\log G)'' \geq 0$, $t \in (0,T)$, and then from inequality (1.48) one deduces

$$G(t) \leq [G(0)]^{1-t/T} [G(T)]^{t/T} . \tag{7.21}$$

Provided we are in the class of solutions such that G is bounded on $[0,T]$, say

$$G(T) \leq M^2,$$

for a known constant M, say, then (7.21) yields a continuous dependence upon the initial data estimate. Indeed, we find recalling (7.3),

$$
\begin{aligned}
G(t) \le & K\Big[< \rho v_i v_i > + T\big\{ < k_{ij} D_{,i} D_{,j} > + < m_{ij} E_{,i} E_{,j} > \\
& + < \ell_{ij} F_{,i} F_{,j} > + < \gamma (D-E)^2 > + < \xi (F-E)^2 > \big\} \Big]^{1-t/T},
\end{aligned}
\tag{7.22}
$$

for t in a compact sub-interval of $[0,T)$.

From (7.9) one sees that the measure G contains $< \rho u_i u_i >$ and so (7.22) yields continuous dependence upon the initial data in the $L^2(\Omega)$ norm of $\mathbf{u}$. In fact, it leads to

$$
\begin{aligned}
& < \rho u_i(\mathbf{x},t) u_i(\mathbf{x},t) > \\
& \quad \le K\Big[< \rho v_i v_i > + T\big\{ < k_{ij} D_{,i} D_{,j} > + < m_{ij} E_{,i} E_{,j} > \\
& \quad + < \ell_{ij} F_{,i} F_{,j} > + < \gamma (D-E)^2 > + < \xi (F-E)^2 > \big\} \Big]^{1-t/T},
\end{aligned}
\tag{7.23}
$$

for $t \in [0,T)$. However, (7.22) does not really represent a useful measure for continuous dependence in p, q or s. One can derive such an estimate but the details are tricky and involve further analysis with the equations (7.1). Details may be calculated following the procedure of Straughan [204] who developed an analogous analysis for the isothermal model of double porosity elasticity.

7.1.2 *Case of* $E(0) > 0$

When $E(0) > 0$ inequality (7.20) still holds. However, we cannot discard the right hand side. Instead, we proceed as in section 1.4.6. Define the functions J and K by

$$
J = G + 2E(0)
$$

and

$$
K = \log J + t^2.
$$

We desire to show $K'' \ge 0$ which is equivalent to

$$
JJ'' - (J')^2 \ge -2J^2. \tag{7.24}
$$

By direct calculation using (7.20) we see that

$$
\begin{aligned}
JJ'' - (J')^2 = & \big(G + 2E(0)\big)G'' - (G')^2 \\
= & GG'' - (G')^2 + 2E(0)G'' \\
\ge & -4E(0)G - 8\big[E(0)\big]^2
\end{aligned}
\tag{7.25}
$$

where in the last line (7.17) has been employed. Upon addition of non-positive terms to the right of (7.25) one therefore finds

$$JJ'' - (J')^2 \geq -8E(0)G - 8\big[E(0)\big]^2 - 2G^2$$
$$= -J^2 . \tag{7.26}$$

Thus, inequality (7.24) holds and so K is a convex function of t in the interval $[0,T)$. Thus we find

$$G(t) + 2E(0) \leq \exp\big[t(T-t)\big]\,\big[M^2 + 2E(0)\big]^{t/T}\,\big[G(0) + 2E(0)\big]^{1-t/T} . \tag{7.27}$$

Inequality (7.27) demonstrates continuous dependence upon the initial data on compact sub-intervals of $[0,T)$, for a solution to the triple porosity problem, in the measure $G(t)$. Given the definition of $G(t)$, (7.9), this also yields continuous dependence in the measure $< \rho u_i u_i >$. Again, extra work is necessary to establish a useful continuous dependence estimate in p, q or s. Further details may be developed following the approach for the double porosity problem in Straughan [204].

7.2 Structural Stability

In the remainder of this chapter we concentrate on structural stability questions. We have already pointed out in section 5.2.1 that structural stability issues, or analysis of continuous dependence upon the model itself, present as great a challenge as continuous dependence upon the initial data, and are equally important.

The precise goal of the remainder of this chapter is to establish that the model for an elastic body with double porosity explained in section 2.3, equations (2.25) - (2.27), is such that the solution depends continuously on changes in the coupling coefficients β_{ij} and γ_{ij} and upon changes in the connectivity coefficient γ. In addition, we show that when β_{ij} and γ_{ij} tend to zero in a certain sense, then the solution converges appropriately.

In the interests of clarity we recall equations (2.25) - (2.27) here and write down the boundary - initial value problem $\mathscr{P}$ for a linear elastic body with a double porosity structure. The boundary - initial value problem $\mathscr{P}$ consists of

$$\begin{aligned}
\rho \ddot{u}_i &= (a_{ijkh} u_{k,h})_{,j} - (\beta_{ij} p)_{,j} - (\gamma_{ij} q)_{,j} \\
\alpha \dot{p} &= (k_{ij} p_{,j})_{,i} - \gamma(p-q) - \beta_{ij} \dot{u}_{i,j} \\
\beta \dot{q} &= (m_{ij} q_{,j})_{,i} + \gamma(p-q) - \gamma_{ij} \dot{u}_{i,j}
\end{aligned} \tag{7.28}$$

in $\Omega \times \{t > 0\}$, together with the boundary conditions

$$u_i(\mathbf{x},t) = u_i^B(\mathbf{x},t), \qquad p(\mathbf{x},t) = p^B(\mathbf{x},t), \qquad q(\mathbf{x},t) = q^B(\mathbf{x},t), \tag{7.29}$$

for $\mathbf{x} \in \Gamma, t > 0$, and the initial conditions

$$\begin{aligned}
u_i(\mathbf{x},0) &= v_i(\mathbf{x}), \qquad \dot{u}_i(\mathbf{x},0) = w_i(\mathbf{x}), \\
p(\mathbf{x},0) &= P(\mathbf{x}), \qquad q(\mathbf{x},0) = Q(\mathbf{x}),
\end{aligned} \tag{7.30}$$

for $\mathbf{x} \in \Omega$, where $u_i^B, p^B, q^B, v_i, w_i, P$ and Q are given functions.

Throughout the remainder of this chapter a_{ijkh} satisfy the symmetries

$$a_{ijkh} = a_{jikh} = a_{khij} \tag{7.31}$$

and a_{ijkh} are positive-definite in the sense that

$$a_{ijkh}\xi_{ij}\xi_{kh} \geq a_0\xi_{ij}\xi_{ij}, \qquad \forall\, \xi_{ij}, \tag{7.32}$$

where $a_0 > 0$ is constant. Furthermore, $k_{ij}, m_{ij}, \beta_{ij}$ and γ_{ij} are symmetric tensors and

$$k_{ij}\xi_i\xi_j \geq k_0\xi_i\xi_i, \qquad m_{ij}\xi_i\xi_j \geq m_0\xi_i\xi_i, \qquad \forall\, \xi_i, \tag{7.33}$$

$k_0 > 0, m_0 > 0$. In addition for $\alpha_0, \beta_0, \rho_0 > 0$, we have

$$\rho(\mathbf{x}) \geq \rho_0, \quad \alpha(\mathbf{x}) \geq \alpha_0, \quad \beta(\mathbf{x}) \geq \beta_0. \tag{7.34}$$

The domain Ω is supposed star shaped with respect to an origin inside.

To commence our analysis of continuous dependence on the coefficients β_{ij} and γ_{ij} we require certain preliminary results.

7.2.1 Auxilliary Functions

Let H be a solution to the problem

$$\begin{aligned} (k_{ij}H_{,j})_{,i} &= 0, \qquad \text{in } \Omega, \\ H &= p^B, \qquad \text{on } \Gamma. \end{aligned} \tag{7.35}$$

Let I be a solution to the problem

$$\begin{aligned} (m_{ij}I_{,j})_{,i} &= 0, \qquad \text{in } \Omega, \\ I &= q^B, \qquad \text{on } \Gamma. \end{aligned} \tag{7.36}$$

Let also G_i be a solution to the problem

$$\begin{aligned} (a_{ijkh}G_{k,h})_{,j} &= 0, \qquad \text{in } \Omega, \\ G_i &= \dot{u}_i^B, \qquad \text{on } \Gamma. \end{aligned} \tag{7.37}$$

When Ω is star shaped with respect to an interior origin, O, Chirita *et al.* [39] have shown by use of Rellich identities, cf. Payne & Weinberger [172], that one has *a priori* bounds for the solutions G_i, H and I, where the bounds depend only on u_i^B, p^B, q^B and the geometry of Ω and Γ. In particular, Chirita *et al.* [39] show that one has truly *a priori* bounds for the quantities

$$< a_{ijkh}G_{i,j}G_{k,h} >, \quad \oint_\Gamma \frac{\partial G_i}{\partial n}\frac{\partial G_i}{\partial n}\,dA \quad \text{and} \quad \|\mathbf{G}\|^2,$$

where $< \cdot >$ denotes integration over Ω and $\|\cdot\|$ is the norm on $L^2(\Omega)$. Chirita *et al.* [39] also show that one has *a priori* bounds for

$$< k_{ij}H_{,i}H_{,j} >, \quad \oint_\Gamma \Big(\frac{\partial H}{\partial n}\Big)^2 dA \quad \text{and} \quad \|H\|^2,$$

and for

$$< m_{ij}I_{,i}I_{,j} >, \quad \oint_\Gamma \Big(\frac{\partial I}{\partial n}\Big)^2 dA \quad \text{and} \quad \|I\|^2.$$

Again, these bounds are truly *a priori* in that they depend only on the data functions p^B and q^B and upon the geometry of the domain Ω. We shall make use of these bounds in this work.

7.2.2 A Priori Estimates

Before proceeding to demonstrate continuous dependence we require some *a priori* estimates for the solution (u_i, p, q) to $\mathscr{P}$ defined in (7.28) - (7.30). Let $(\cdot,\cdot)$ denote the inner product on $L^2(\Omega)$ and then with (u_i, p, q) denoting a solution to (7.28) - (7.30) and H, I and G_i being the solutions to (7.35) - (7.37) we form the combinations

$$\int_0^t \Big((a_{ijkh}u_{k,h})_{,j} - (\beta_{ij}p)_{,j} - (\gamma_{ij}q)_{,j} - \rho\ddot{u}_i\,, G_i - \dot{u}_i\Big)ds = 0, \tag{7.38}$$

and

$$\int_0^t \Big((k_{ij}p_{,j})_{,i} - \gamma(p-q) - \beta_{ij}\dot{u}_{i,j} - \alpha\dot{p}, H - p\Big)ds = 0, \tag{7.39}$$

together with

$$\int_0^t \Big((m_{ij}q_{,j})_{,i} + \gamma(p-q) - \gamma_{ij}\dot{u}_{i,j} - \beta\dot{q}, I - q\Big)ds = 0. \tag{7.40}$$

We now wish to utilize these relations to yield suitable *a priori* estimates for (u_i, p, q) and this is achieved by performing various integrations by parts on each of these equations recalling the boundary conditions on H, I and G_i as given by (7.35) - (7.37).

From equation (7.39) we integrate the $(k_{ij}p_{,j})_{,i}$ term recalling $H = p^B$ on Γ and then one may show

$$\begin{aligned}
&\int_0^t < k_{ij}p_{,i}p_{,j} > ds - \int_0^t < k_{ij}p_{,j}H_{,i} > ds + \int_0^t < \gamma p(p-q) > ds \\
&- \int_0^t < \gamma H(p-q) > ds - < \alpha Hp > + < \alpha H(0)p(0) > \\
&+ \int_0^t < \alpha \dot{H} p > ds + \frac{1}{2} < \alpha p^2 > - \frac{1}{2} < \alpha p^2(0) > \\
&+ \int_0^t < \beta_{ij} p \dot{u}_{i,j} > ds - \int_0^t \oint_\Gamma n_j \dot{u}_i \beta_{i,j} H \, dA \, ds \\
&+ \int_0^t < \beta_{ij} \dot{u}_i H_{,j} > ds = 0,
\end{aligned} \tag{7.41}$$

where $< \cdot >$ denotes integration over Ω, $H(0) = H(\mathbf{x},0)$, $p(0) = p(\mathbf{x},0)$, and we have deliberately left the third last term on the left of (7.41), involving $\dot{u}_{i,j}$, to balance with a term which will arise from (7.38).

The second term in (7.41) is handled as follows

$$\begin{aligned}
\int_0^t < k_{ij}p_{,j}H_{,i} > ds &= \int_0^t \oint_\Gamma n_j k_{ij} p H_{,i} \, dA \, ds - \int_0^t < (k_{ij}H_{,i})_{,j} p > ds \\
&= \int_0^t \oint_\Gamma n_j k_{ij} p H_{,i} \, dA \, ds,
\end{aligned} \tag{7.42}$$

where we have employed (7.35).

A similar rearrangement is executed on (7.40) to arrive at analogous equations to (7.41) and (7.42). We then add the results to form a single equation. However, the terms involving γ may be arranged to have

$$\int_0^t < \gamma(p-q)^2 > ds$$

on the left hand side while on the right one has

$$I_1 = \int_0^t < \gamma H(p-q) > ds - \int_0^t < \gamma I(p-q) > ds.$$

The arithmetic-geometric mean inequality is applied to these two terms to see that

$$I_1 \leq \int_0^t \frac{1}{2} < \gamma(H^2 + I^2) > ds + \int_0^t < \gamma(p-q)^2 > ds.$$

In this way we may remove the $\gamma(p-q)^2$ terms and there remains the $\gamma(H^2+I^2)$ contribution.

In this manner one may deduce from (7.41), (7.42), and the equivalent inequalities arising from (7.40),

$$
\begin{aligned}
&\int_0^t < k_{ij}p_{,i}p_{,j} > ds + \int_0^t < m_{ij}q_{,i}q_{,j} > ds + \frac{1}{2} < \alpha p^2 > \\
&\quad + \frac{1}{2} < \beta q^2 > + \int_0^t < \gamma_{ij}\dot{u}_{i,j}q > ds + \int_0^t < \beta_{ij}\dot{u}_{i,j}p > ds \\
&\quad \le \int_0^t \oint_\Gamma n_j k_{ij} p^B H_{,i}\, dA\, ds + \int_0^t \oint_\Gamma n_j m_{ij} q^B I_{,i}\, dA\, ds \\
&\quad + \int_0^t \oint_\Gamma n_j \gamma_{ij} \dot{u}_i^B I\, dA\, ds + \int_0^t \oint_\Gamma n_j \beta_{ij} \dot{u}_i^B H\, dA\, ds \\
&\quad - \int_0^t < \gamma_{ij}\dot{u}_i I_{,j} > ds - \int_0^t < \beta_{ij}\dot{u}_i H_{,j} > ds \\
&\quad + < \beta I q > + < \alpha H p > - \int_0^t < \beta q \dot{I} > ds - \int_0^t < \alpha p \dot{H} > ds \\
&\quad - < \alpha H(0) P > - < \beta I(0) Q > + \frac{1}{2} < \beta Q^2 > \\
&\quad + \frac{1}{2} < \alpha P^2 > + \frac{1}{2}\int_0^t < \gamma(H^2 + I^2) > ds.
\end{aligned} \tag{7.43}
$$

After some integration by parts on (7.38) and use of (7.37) one may arrive at

$$
\begin{aligned}
&\frac{1}{2} < a_{ijkh}u_{i,j}u_{k,h} > + \frac{1}{2} < \rho \dot{u}_i \dot{u}_i > - \int_0^t < \gamma_{ij} q \dot{u}_{i,j} > ds \\
&\quad - \int_0^t < \beta_{ij} p \dot{u}_{i,j} > ds = \frac{1}{2} < a_{ijkh} v_{i,j} v_{k,h} > + \frac{1}{2} < \rho w_i w_i > \\
&\quad - < \rho w_i G(0) > - \int_0^t < \rho \dot{u}_i G > ds + < \rho \dot{u}_i G > \\
&\quad - \int_0^t < \gamma_{ij} q G_{i,j} > ds - \int_0^t < \beta_{ij} p G_{i,j} > ds \\
&\quad + \int_0^t < a_{ijkh} u_{k,h} G_{i,j} > ds,
\end{aligned} \tag{7.44}
$$

where (7.30) have been employed.

The last term in (7.44) is further integrated to find

$$
\begin{aligned}
\int_0^t < a_{ijkh} u_{k,h} G_{i,j} > ds &= \int_0^t \oint_\Gamma a_{ijkh} u_k n_h G_{i,j}\, dA\, ds \\
&\quad - \int_0^t < (a_{ijkh} G_{i,j})_{,h} u_k > ds \\
&= \int_0^t \oint_\Gamma a_{ijkh} u_k n_h G_{i,j}\, dA\, ds,
\end{aligned} \tag{7.45}
$$

where in deriving (7.45), (7.37) has been utilized.

We now combine (7.43) with (7.44) and (7.45) to deduce the inequality

$$\begin{aligned}\int_0^t <k_{ij}p_{,i}p_{,j}> ds + \int_0^t < m_{ij}q_{,i}q_{,j} > ds + \frac{1}{2} < \alpha p^2 > \\ + \frac{1}{2} < \beta q^2 > + \frac{1}{2} < a_{ijkh}u_{i,j}u_{k,h} > + \frac{1}{2} < \rho \dot{u}_i\dot{u}_i > \\ \leq I_B + I_D + I_S + I_T ,\end{aligned} \tag{7.46}$$

where the terms I_B, I_D, I_S and I_T are given below. The term I_B involves boundary integrals and is

$$\begin{aligned}I_B = & \int_0^t \oint_\Gamma a_{ijkh}u_k n_h G_{i,j}\, dA\, ds + \int_0^t \oint_\Gamma n_j k_{ij} p^B H_{,i}\, dA\, ds \\ & + \int_0^t \oint_\Gamma n_j m_{ij} q^B I_{,i}\, dA\, ds + \int_0^t \oint_\Gamma n_j \gamma_{ij} \dot{u}_i I\, dA\, ds \\ & + \int_0^t \oint_\Gamma n_j \beta_{ij} \dot{u}_i H\, dA\, ds.\end{aligned} \tag{7.47}$$

The term I_D involves initial data and is

$$\begin{aligned}I_D = & \frac{1}{2} < a_{ijkh}v_{i,j}v_{k,h} > + \frac{1}{2} < \rho w_i w_i > - < \alpha H(0) P > \\ & - < \beta I(0) Q > - < \rho w_i G(0) > + \frac{1}{2} < \beta Q^2 > + \frac{1}{2} < \alpha P^2 > .\end{aligned} \tag{7.48}$$

The terms I_S and I_T involve the solution and are given by

$$I_S = < \beta I q > + < \alpha H p > + < \rho \dot{u}_i G >, \tag{7.49}$$

and

$$\begin{aligned}I_T = & \frac{1}{2} \int_0^t < \gamma (H^2 + I^2) > ds \\ & - \int_0^t < \gamma_{ij} \dot{u}_i I_{,j} > ds - \int_0^t < \beta_{ij} \dot{u}_i H_{,j} > ds \\ & - \int_0^t < \gamma_{ij} q G_{i,j} > ds - \int_0^t < \beta_{ij} p G_{i,j} > ds \\ & - \int_0^t < \rho \dot{u}_i \dot{G} > ds - \int_0^t < \beta q \dot{I} > ds - \int_0^t < \alpha p \dot{H} > ds.\end{aligned} \tag{7.50}$$

We remark that the five boundary terms comprising I_B may be regarded as data terms. To see this we recall that on Γ we may write $H_{,i}$ as

$$H_{,i} = n_i \frac{\partial H}{\partial n} + a^{\alpha\beta} x^i_{;\alpha} H_{;\beta}$$

where n_i is the unit outward normal to Γ; $\partial H / \partial n$ is the normal derivative on Γ; s^α, $\alpha = 1,2$ are surface coordinates on Γ, and $a^{\alpha\beta}$ is the fundamental form for the surface. Since we have bounds for $\int_\Gamma (\partial H / \partial n)^2 dA$ and $H = p^B$ on Γ such terms as the second on the right of (7.47) may be written as

$$\int_0^t \oint_\Gamma n_j k_{ij} p H_{,i}\, dA\, ds = \int_0^t \oint_\Gamma k_{ij} p^B n_i n_j \frac{\partial H}{\partial n}\, dA\, ds + \int_0^t \oint_\Gamma k_{ij} p^B n_j a^{\alpha\beta} x^i_{;\beta} H_{;\alpha}\, dAds$$

and these quantities may all be bounded by known values involving boundary data or the geometry of Ω, Γ.

Suppose now the cofficients are bounded in the sense that

$$|\gamma_{ij}|, |\beta_{ij}|, |\gamma|, |\beta|, |\alpha| \leq M,$$

for some number M. Then, after extensive use of the arithmetic-geometric mean inequality on the right of (7.46) one may arrive at

$$\begin{aligned} &\int_0^t <k_{ij} p_{,i} p_{,j}> ds + \int_0^t < m_{ij} q_{,i} q_{,j} > ds + \frac{1}{4} < \alpha p^2 > \\ &+ \frac{1}{4} < \beta q^2 > + \frac{1}{2} < a_{ijkh} u_{i,j} u_{k,h} > + \frac{1}{4} < \rho \dot{u}_i \dot{u}_i > \\ &\leq L_0 + L_B + \Big(\frac{M}{\rho_0} + \frac{1}{2}\Big) \int_0^t < \rho \dot{u}_i \dot{u}_i > ds \\ &+ \frac{M}{\beta_0} \int_0^t < \beta q^2 > ds + \frac{M}{\alpha_0} \int_0^t < \alpha p^2 > ds + \mathscr{F}, \end{aligned} \tag{7.51}$$

where the terms L_0, L_B and $\mathscr{F}$ are given below.

$$\begin{aligned} L_0 =& \frac{1}{2} < a_{ijkh} v_{i,j} v_{k,h} > + \frac{1}{2} < \rho w_i w_i > - < \alpha H(0) P > \\ &- < \beta I(0) Q > - < \rho w_i G(0) > + \frac{1}{2} < \beta Q^2 > + \frac{1}{2} < \alpha P^2 >, \end{aligned} \tag{7.52}$$

and

$$\begin{aligned} L_B =& \int_0^t \oint_\Gamma a_{ijkh} n_h u_k^B G_{i,j}\, dA\, ds + \int_0^t \oint_\Gamma n_j k_{ij} p^B H_{,i}\, dA\, ds \\ &+ \int_0^t \oint_\Gamma n_j m_{ij} q^B I_{,i}\, dA\, ds + \int_0^t \oint_\Gamma n_j \gamma_{ij} \dot{u}_i^B I\, dA\, ds \\ &+ \int_0^t \oint_\Gamma n_j \beta_{ij} \dot{u}_i^B H\, dA\, ds \end{aligned} \tag{7.53}$$

and additionally

$$\begin{aligned} \mathscr{F} =& \frac{M}{2} \int_0^t \|\nabla I\|^2 ds + \frac{M}{2} \int_0^t \|\nabla H\|^2 ds + M \int_0^t \|\nabla \mathbf{G}\|^2 ds \\ &+ \frac{M}{2} \int_0^t \|\mathbf{G}\|^2 ds + \frac{M}{2} \int_0^t \|\dot{I}\|^2 ds + \frac{M}{2} \int_0^t \|\dot{H}\|^2 ds \\ &+ M\|\mathbf{G}\|^2 + \frac{3M}{2} (\|I\|^2 + \|H\|^2). \end{aligned} \tag{7.54}$$

The functions $v_i, w_i, P, Q, u_i^B, p^B, q^B$ are data, and as pointed out in section 7.2.1, the terms

$$
\begin{aligned}
&\|\nabla \mathbf{G}\|^2, \quad \|\mathbf{G}\|^2, \quad \oint_\Gamma (\partial G_i/\partial n)(\partial G_i/\partial n)dA, \\
&\|\nabla H\|^2, \quad \|H\|^2, \quad \oint_\Gamma (\partial H/\partial n)^2\,dA, \\
&\|\nabla I\|^2, \quad \|I\|^2 \quad \text{and} \quad \oint_\Gamma (\partial I/\partial n)^2\,dA
\end{aligned}
$$

are also bounded by data. Hence, the terms L_0, L_B and $\mathscr{F}$ are all boundable by truly *a priori* estimates, i.e. by known quantities.

From (7.51) one may derive

$$
\dot{F} \le D + KF \tag{7.55}
$$

where

$$
F = \int_0^t (<\alpha p^2> + <\beta q^2> + <\rho \dot{u}_i \dot{u}_i>)ds,
$$

and

$$
K = \max\left\{\frac{M}{\rho_0} + \frac{1}{2}, \frac{M}{\beta_0}, \frac{M}{\alpha_0}\right\},
$$

and where D is effectively a data term given by

$$
D = 4(L_0 + L_B + \mathscr{F}).
$$

Inequality (7.55) is integrated to find

$$
\int_0^t (<\alpha p^2> + <\beta q^2> + <\rho \dot{u}_i \dot{u}_i>)ds \le \frac{D}{K} e^{Kt}. \tag{7.56}
$$

Further, employing (7.55) in (7.56) we find

$$
<\alpha p^2> + <\beta q^2> + <\rho \dot{u}_i \dot{u}_i> \le D + De^{Kt}. \tag{7.57}
$$

Inequality (7.57) represents an *a priori* inequality for p, q and $\dot{u}_i$ in the weighted norms as indicated, but thanks to (7.34) in the $L^2(\Omega)$ norm also.

If we now return to inequality (7.51) then we may see that

$$
\begin{aligned}
\int_0^t <k_{ij}p_{,i}p_{,j}> ds + \int_0^t <m_{ij}q_{,i}q_{,j}> ds + \frac{a_0}{2}\|\nabla \mathbf{u}\|^2 \\
\le L_0 + L_B + \mathscr{F} + KF,
\end{aligned} \tag{7.58}
$$

and so by employing also (7.56) we obtain *a priori* bounds for the terms

$$
\int_0^t \|\nabla p\|^2 ds, \quad \int_0^t \|\nabla q\|^2 ds \quad \text{and} \quad \|\nabla \mathbf{u}\|^2.
$$

7.2.3 Continuous Dependence on β_{ij} (or γ_{ij})

We now establish continuous dependence on β_{ij} (or γ_{ij}) for a solution to the boundary-initial value problem $\mathscr{P}$ given by (7.28) - (7.30). It is sufficient to establish continuous dependence on β_{ij} since it is clear from the mathematical structure of equations (7.28) that the calculation for continuous dependence upon γ_{ij} is exactly the same as that for β_{ij} excepting one swaps q for p. Hence, we only include details for β_{ij}.

Let (u_i^1, p^1, q^1) be a solution which satisfies $\mathscr{P}$ for functions $u_i^B, p^B, q^B, v_i, w_i, P$ and Q, and for a coefficient β_{ij}^1. Let (u_i^2, p^2) be another solution to $\mathscr{P}$ for the same functions $u_i^B, p^B, q^B, v_i, w_i, P$ and Q, but for a different coefficient β_{ij}, say β_{ij}^2. The other coeffcients in (7.28) are the same for both solutions.

Define the difference variables by

$$u_i = u_i^1 - u_i^2, \quad p = p^1 - p^2, \quad q = q^1 - q^2, \quad \beta_{ij} = \beta_{ij}^1 - \beta_{ij}^2. \tag{7.59}$$

Then one may verify that (u_i, p, q) satisfies the boundary-initial value problem

$$\begin{aligned}
\rho \ddot{u}_i &= (a_{ijkh} u_{k,h})_{,j} - (\gamma_{ij} q)_{,j} - (\beta_{ij}^1 p)_{,j} - (\beta_{ij} p^2)_{,j}, \\
\alpha \dot{p} &= (k_{ij} p_{,j})_{,i} - \gamma(p-q) - \beta_{ij}^1 \dot{u}_{i,j} - \beta_{ij} \dot{u}_{i,j}^2, \\
\beta \dot{q} &= (m_{ij} q_{,j})_{,i} + \gamma(p-q) - \gamma_{ij} \dot{u}_{i,j},
\end{aligned} \tag{7.60}$$

in $\Omega \times \{t > 0\}$, together with the boundary conditions

$$u_i(\mathbf{x},t) = 0, \qquad p(\mathbf{x},t) = 0, \qquad q(\mathbf{x},t) = 0, \qquad \text{on } \Gamma \times \{t > 0\}, \tag{7.61}$$

and the initial conditions

$$u_i(\mathbf{x},0) = 0, \quad \dot{u}_i(\mathbf{x},0) = 0, \quad p(\mathbf{x},0) = 0, \quad q(\mathbf{x},0) = 0, \qquad \mathbf{x} \in \Omega. \tag{7.62}$$

Multiply equation $(7.60)_1$ by $\dot{u}_i$, equation $(7.60)_2$ by p, equation $(7.60)_3$ by q and integrate each over Ω. One may obtain

$$\begin{aligned}
&\frac{d}{dt}\Big[\frac{1}{2} < \rho \dot{u}_i \dot{u}_i > + \frac{1}{2} < a_{ijkh} u_{i,j} u_{k,h} > \Big] - < \gamma_{ij} q \dot{u}_{i,j} > \\
&\quad - < \beta_{ij}^1 p \dot{u}_{i,j} > = - < \beta_{ij} p_{,j}^2 \dot{u}_i > - < \beta_{ij,j} p^2 \dot{u}_i >,
\end{aligned} \tag{7.63}$$

together with

$$\begin{aligned}
&\frac{d}{dt}\Big[\frac{1}{2} < \alpha p^2 > + < k_{ij} p_{,i} p_{,j} > \Big] + < \gamma p (p-q) > \\
&\quad + < \beta_{ij}^1 p \dot{u}_{i,j} > = < \beta_{ij} p_{,j} \dot{u}_i^2 > + < \beta_{ij,j} p \dot{u}_i^2 >,
\end{aligned} \tag{7.64}$$

and

$$\frac{d}{dt}\Big[\frac{1}{2}<\beta q^2>+<m_{ij}q_{,i}q_{,j}>\Big]-<\gamma q(p-q)> \\ +<\gamma_{ij}q\dot{u}_{i,j}>=0. \tag{7.65}$$

We next add (7.63), (7.64) and (7.65) and then define β_m^2 and β_M^2 by

$$\beta_m^2=\beta_{ij}\beta_{ij}, \qquad \beta_M^2=\beta_{ij,j}\beta_{ik,k}. \tag{7.66}$$

We employ the arithmetic-geometric mean inequality as follows,

$$\begin{aligned} |<\beta_{ij}p_{,j}^2\dot{u}_i>| &\le \frac{\alpha_1}{2}\beta_m^2\|\nabla p\|^2+\frac{1}{2\alpha_1\rho_0}<\rho\dot{u}_i\dot{u}_i>,\\ |<\beta_{ij,j}p^2\dot{u}_i>| &\le \frac{\alpha_2}{2}\beta_M^2\|p^2\|^2+\frac{1}{2\alpha_2\rho_0}<\rho\dot{u}_i\dot{u}_i>, \end{aligned} \tag{7.67}$$

and

$$\begin{aligned} |<\beta_{ij}p_{,j}\dot{u}_i^2>| &\le \frac{\alpha_3}{2}\beta_m^2\|\dot{\mathbf{u}}\|^2+\frac{1}{2\alpha_3 k_0}<k_{ij}p_{,i}p_{,j}>,\\ |<\beta_{ij,j}p\dot{u}_i^2>| &\le \frac{\alpha_4}{2}\beta_M^2\|\dot{\mathbf{u}}\|^2+\frac{1}{2\alpha_4\alpha_0}<\alpha p^2>, \end{aligned} \tag{7.68}$$

for constants $\alpha_1,\ldots,\alpha_4$ to be selected.

Choose $\alpha_1=1/2\rho_0, \alpha_2=1/2\rho_0$, $\alpha_3=1/2k_0, \alpha_4=1/\alpha_0$, and then employing (7.67) and (7.68) in the sum of (7.63) - (7.65) one may derive

$$\begin{aligned} &\frac{d}{dt}\Big[<\rho\dot{u}_i\dot{u}_i>+<a_{ijkh}u_{i,j}u_{k,h}>+<\alpha p^2>+<\beta q^2>\Big]\\ &\le<\alpha p^2>+<\rho\dot{u}_i\dot{u}_i>+\frac{\beta_M^2}{2\rho_0}\|p^2\|^2\\ &\quad+\Big(\frac{\beta_M^2}{\alpha_0}+\frac{\beta_m^2}{2k_0}\Big)\|\dot{\mathbf{u}}\|^2+\frac{\beta_m^2}{2\rho_0}\|\nabla p^2\|^2. \end{aligned} \tag{7.69}$$

Now $\|p^2\|$ and $\|\dot{\mathbf{u}}^2\|$ are bounded by *a priori* constants which depend only on data and on the geometry of Ω, hence we put

$$F=\int_0^t\Big[<\rho\dot{u}_i\dot{u}_i>+<a_{ijkh}u_{i,j}u_{k,h}>+<\alpha p^2>+<\beta q^2>\Big]ds$$

to derive from (7.69)

$$F'\le F+k_1\beta_M^2+k_2\beta_m^2+\frac{\beta_m^2}{2\rho_0}\int_0^t\|\nabla p^2\|^2ds, \tag{7.70}$$

where

$$k_1=\Big(\frac{K_1^2}{2\rho_0}+\frac{K_2^2}{\alpha_0}\Big)T, \qquad k_2=\frac{TK_2^2}{2k_0},$$

where K_1 and K_2 denote *a priori* bounds for $\|p^2\|$ and $\|\dot{\mathbf{u}}\|^2$, respectively, and $T < \infty$ represents a time interval with $t < T$. Since we have an *a priori* bound for $\int_0^t \|\nabla p^2\|^2 ds$ we may now conclude from (7.70) that

$$F' \leq F + D, \tag{7.71}$$

where $D = k_1\beta_M^2 + (k_2 + K_3^2/2\rho_0)\beta_m^2$, K_3^2 being an *a priori* bound for $\int_0^t \|\nabla p^2\|^2 ds$. One integrates (7.71) to find

$$F(t) \leq TDe^T, \tag{7.72}$$

for $0 < t \leq T$. Hence, from (7.72) we find there are constants c_1 and c_2 depending only on T, the data of the problem and the geometry of Ω such that

$$F(t) \leq c_1\beta_m^2 + c_2\beta_M^2. \tag{7.73}$$

An appeal to (7.71) then allows us to see there are *a priori* constants c_3 and c_4 such that

$$< \rho\dot{u}_i\dot{u}_i > + < a_{ijkh}u_{i,j}u_{k,h} > + < \alpha p^2 > + < \beta q^2 > \leq c_3\beta_m^2 + c_4\beta_M^2, \tag{7.74}$$

for $0 < t \leq T$. Inequality (7.74) demonstrates continuous dependence of the solution to $\mathscr{P}$ upon the coupling coefficient β_{ij}.

7.2.4 Continuous Dependence on γ

We now investigate continuous dependence upon the connectivity coefficient γ for the solution to the boundary-initial value problem $\mathscr{P}$ given by (7.28) - (7.30). Thus, we let (u_i^1, p^1, q^1) and (u_i^2, p^2, q^2) be solutions to $\mathscr{P}$ for the same values of $u_i^B, p^B, q^B, v_i, w_i, P$ and Q, and for the same coefficients $\rho, \alpha, \beta, a_{ijkh}, \beta_{ij}, \gamma_{ij}, k_{ij}$ and m_{ij}. However, (u_i^1, p^1, q^1) satisfies $\mathscr{P}$ for a gamma with value γ_1 whereas (u_i^2, p^2, q^2) is a solution to $\mathscr{P}$ with γ value of γ_2. The difference variables u_i, p, q and γ are defined by

$$u_i = u_i^1 - u_i^2, \quad p = p^1 - p^2, \quad q = q^1 - q^2, \quad \gamma = \gamma_1 - \gamma_2.$$

By subtraction one shows that (u_i, p, q) satisfies the boundary-initial value problem given by

$$\begin{aligned} \rho\ddot{u}_i &= (a_{ijkh}u_{k,h})_{,j} - (\beta_{ij}p)_{,j} - (\gamma_{ij}q)_{,j}, \\ \alpha\dot{p} &= (k_{ij}p_{,j})_{,i} - \gamma(p^1 - q^1) - \gamma_2(p - q) - \beta_{ij}\dot{u}_{i,j}, \\ \beta\dot{q} &= (m_{ij}q_{,j})_{,i} + \gamma(p^1 - q^1) + \gamma_2(p - q) - \gamma_{ij}\dot{u}_{i,j}, \end{aligned} \tag{7.75}$$

in $\Omega \times \{t > 0\}$, together with the boundary conditions

$$u_i(\mathbf{x}, t) = 0, \qquad p(\mathbf{x}, t) = 0, \qquad q(\mathbf{x}, t) = 0, \qquad \text{on } \Gamma \times \{t > 0\}, \tag{7.76}$$

and the initial conditions

$$u_i(\mathbf{x},0)=0, \quad \dot{u}_i(\mathbf{x},0)=0, \quad p(\mathbf{x},0)=0, \quad q(\mathbf{x},0)=0, \qquad \mathbf{x}\in\Omega. \tag{7.77}$$

The coefficients in (7.75) are required to satisfy the conditions in (7.31) - (7.34) and

$$\gamma_2(\mathbf{x})\geq\gamma_0>0 \tag{7.78}$$

for a constant γ_0.

The analysis commences by multiplying $(7.75)_1$ by $\dot{u}_i$, $(7.75)_2$ by p, $(7.75)_3$ by q, integrating each over Ω, and after summing the results and some integration by parts one may show that

$$\begin{aligned}&\frac{1}{2}\frac{d}{dt}\Big[<\rho\dot{u}_i\dot{u}_i>+<a_{ijkh}u_{i,j}u_{k,h}>+<\alpha p^2>+<\beta q^2>\Big] \qquad (7.79)\\&\quad+<k_{ij}p_{,i}p_{,j}>+<m_{ij}q_{,i}q_{,j}>+<\gamma_2(p-q)^2>\\&\quad=<\gamma(p^1-q^1)(p-q)>\\&\quad\leq<\frac{\gamma^2}{4\gamma_2}(p^1-q^1)^2>+<\gamma_2(p-q)^2>, \qquad (7.80)\end{aligned}$$

where in the last line we have employed the arithmetic-geometric mean inequality.

Thus, from (7.80) we deduce

$$\begin{aligned}&\frac{1}{2}\frac{d}{dt}\Big[<\rho\dot{u}_i\dot{u}_i>+<a_{ijkh}u_{i,j}u_{k,h}>+<\alpha p^2>+<\beta q^2>\Big]\\&\quad+<k_{ij}p_{,i}p_{,j}>+<m_{ij}q_{,i}q_{,j}>\\&\quad\leq\frac{\|p^1-q^1\|^2}{4\gamma_0}\gamma_m^2,\end{aligned} \tag{7.81}$$

where $\gamma_m=\max_\Omega|\gamma|$ and γ_0 is given in (7.78).

From section 7.2.2 we know that $\|p^1\|$ and $\|q^1\|$ are bounded by a constant depending only on data and the geometry of Ω. Thus, using the triangle inequality the right hand side of (7.81) is bounded by a term of form $c_1\gamma_m^2$ where c_1 is an *a priori* known constant. Thus, from (7.81) we may derive a continuous dependence upon γ estimate. This is of the form

$$\begin{aligned}&\frac{1}{2}\big[<\rho\dot{u}_i\dot{u}_i>+<a_{ijkh}u_{i,j}u_{k,h}>+<\alpha p^2>+<\beta q^2>\big]\\&\quad+\int_0^t<k_{ij}p_{,i}p_{,j}>ds+\int_0^t<m_{ij}q_{,i}q_{,j}>ds\\&\quad\leq c_1T\gamma_m^2,\end{aligned} \tag{7.82}$$

for $0<t\leq T$, some $T<\infty$. This inequality also yields continuous dependence upon γ for $\|\mathbf{u}\|, \|\nabla\mathbf{u}\|, \|p\|, \|q\|, \int_0^t\|\nabla p\|^2ds$ and $\int_0^t\|\nabla q\|^2ds$.

7.2.5 Convergence

The goal of this section is to study how a solution to $\mathscr{P}$ given by (7.28) - (7.30) converges as $\beta_{ij} \to 0, \gamma_{ij} \to 0$ in a precise sense. We show that under precise conditions the solution will converge to a solution to the equations of linear elastodynamics together with a solution to the Barenblatt & Zheltov [12] system of section 2.1.

Thus, let (u_i^1, p^1, q^1) be a solution to (7.28) - (7.30) and let (u_i^2, p^2, q^2) be a solution to the same equations and the same boundary and initial conditions but with $\beta_{ij} = 0, \gamma_{ij} = 0$. Define the difference solution (u_i, p, q) as in (7.59) and then one may verify that this solution satisfies the boundary-initial value problem

$$\begin{aligned}
\rho \ddot{u}_i &= (a_{ijkh} u_{k,h})_{,j} - (\beta_{ij} p^1)_{,j} - (\gamma_{ij} q^1)_{,j}, \\
\alpha \dot{p} &= (k_{ij} p_{,j})_{,i} - \gamma(p-q) - \beta_{ij} \dot{u}^1_{i,j}, \\
\beta \dot{q} &= (m_{ij} q_{,j})_{,i} + \gamma(p-q) - \gamma_{ij} \dot{u}^1_{i,j},
\end{aligned} \tag{7.83}$$

in $\Omega \times \{t > 0\}$, together with the same zero boundary and initial conditions as (7.76) and (7.77).

The analysis of convergence commences by multiplying equation $(7.83)_1$ by $\dot{u}_i$, $(7.83)_2$ by p, and $(7.83)_3$ by q, and then integrating each in turn over Ω. After some integration by parts and adding the three resulting equations one may arrive at

$$\begin{aligned}
&\frac{1}{2} \frac{d}{dt} \Big[< \rho \dot{u}_i \dot{u}_i > + < a_{ijkh} u_{i,j} u_{k,h} > + < \alpha p^2 > + < \beta q^2 > \Big] \\
&\quad + < k_{ij} p_{,i} p_{,j} > + < m_{ij} q_{,i} q_{,j} > + < \gamma (p-q)^2 > \\
&\quad = - < \beta_{ij} \dot{u}^1_{i,j} p > - < \gamma_{ij} \dot{u}^1_{i,j} q > \\
&\qquad - < (\beta_{ij} p^1)_{,j} \dot{u}_i > - < (\gamma_{ij} q^1)_{,j} \dot{u}_i > \\
&\quad =< \beta_{ij} p_{,j} \dot{u}^1_i > + < \gamma_{ij} q_{,j} \dot{u}^1_i > + \beta_{ij,j} p \dot{u}^1_i > \\
&\qquad + < \gamma_{ij,j} q \dot{u}^1_i > - < \beta_{ij} p^1_{,j} \dot{u}_i > - < \beta_{ij,j} p^1 \dot{u}_i > \\
&\qquad - < \gamma_{ij} q^1_{,j} \dot{u}_i > - < \gamma_{ij,j} q^1 \dot{u}_i > .
\end{aligned} \tag{7.84}$$

We handle the first two terms on the right of (7.84) by using the arithmetic-geometric mean inequality as follows

$$\begin{aligned}
&< \beta_{ij} p_{,j} \dot{u}^1_i > + < \gamma_{ij} q_{,j} \dot{u}^1_i > \\
&\quad \le \frac{1}{2\alpha_1} \|\nabla p\|^2 + \frac{1}{2\alpha_2} \|\nabla q\|^2 + \beta_M^2 \Big(\frac{\alpha_1 + \alpha_2}{2} \Big) \|\dot{\mathbf{u}}^1\|^2 \\
&\quad \le \frac{1}{2k_0 \alpha_1} < k_{ij} p_{,i} p_{,j} > + \frac{1}{2m_0 \alpha_2} < m_{ij} q_{,i} q_{,j} > \\
&\qquad + \beta_M^2 \Big(\frac{\alpha_1 + \alpha_2}{2} \Big) \|\dot{\mathbf{u}}^1\|^2,
\end{aligned} \tag{7.85}$$

where $\alpha_1, \alpha_2 > 0$ are to be selected and

$$\beta_M^2 = \max\left\{\max_{\Omega} |\beta_{ij}|, \max_{\Omega} |\gamma_{ij}|\right\}.$$

Pick $\alpha_1 = 1/k_0, \alpha_2 = 1/m_0$, and then employ (7.85) in (7.84) to find, after judicious use of the arithmetic-geometric mean inequality

$$\begin{aligned}
&\frac{1}{2}\frac{d}{dt}\Big[< \rho\dot{u}_i\dot{u}_i > + < a_{ijkh}u_{i,j}u_{k,h} > + < \alpha p^2 > + < \beta q^2 >\Big] \\
&\quad + \frac{1}{2} < k_{ij}p_{,i}p_{,j} > + \frac{1}{2} < m_{ij}q_{,i}q_{,j} > + < \gamma(p-q)^2 > \\
&\quad \leq \frac{2}{\rho_0}\big[< \rho\dot{u}_i\dot{u}_i > + < \alpha p^2 > + < \beta q^2 >\big] \\
&\qquad + (k_1+k_2)\beta_M^2 + k_3 B_M^2 ,
\end{aligned} \tag{7.86}$$

where

$$B_M^2 = \max\left\{\max_{\Omega} |\beta_{ij,j}|, \max_{\Omega} |\gamma_{ij,j}|\right\}$$

and

$$\begin{aligned}
k_1 &= \frac{1}{2}\Big(\frac{1}{k_0} + \frac{1}{m_0}\Big)\|\dot{\mathbf{u}}^1\|^2, \qquad k_2 = \frac{1}{2}\|\nabla p^1\|^2 + \frac{1}{2}\|\nabla q^1\|^2, \\
k_3 &= \frac{\rho_0}{8}\Big(\frac{1}{\alpha_0} + \frac{1}{\beta_0}\Big)\|\dot{\mathbf{u}}^1\|^2 + \frac{1}{2}(\|p^1\|^2 + \|q^1\|^2).
\end{aligned}$$

From section 7.2.2 we know that $\|\dot{\mathbf{u}}^1\|$, $\|p^1\|$ and $\|q^1\|$ are bounded by *a priori* constants and so $k_1 \leq K_1$, $k_3 \leq K_3$ for constants K_1 and K_3.

Define E and F by

$$E(t) = < \rho\dot{u}_i\dot{u}_i > + < a_{ijkh}u_{i,j}u_{k,h} > + < \alpha p^2 > + < \beta q^2 >,$$

and

$$F(t) = \int_0^t E(s)ds,$$

and then by integrating (7.86) over $(0,t)$ and dropping some positive terms one may show, for $t \in (0,T)$,

$$E \leq K_1 T\beta_M^2 + K_3 T B_M^2 + \beta_M^2\Big(\int_0^t k_2 ds\Big) + K\int_0^t E\,ds, \tag{7.87}$$

where $K = 4/\rho_0$. From section 7.2.2 we also know $\int_0^t k_2 ds$ is *a priori* bounded by a known constant K_4 say. Thus, put $K_2 = K_1T + K_4$ and then from (7.87) we may deduce

$$F' - KF \leq B, \tag{7.88}$$

where

$$B = K_2\beta_M^2 + K_3 T B_M^2 .$$

Upon integration of (7.88) we obtain

$$F(t) \leq \frac{B}{K} e^{Kt}$$

and then by employing (7.88) and (7.87) it follows

$$E(t) \leq B + KF \leq B(1 + e^{Kt}). \tag{7.89}$$

Recalling the definition of $E(t)$ one sees that inequality (7.89) shows that as $\beta_{ij}, \gamma_{ij} \to 0$ in the sense that $B \to 0$ then $E(t) \to 0$ and the solution converges to that of linear elastodynamics and to the Barenblatt & Zheltov equations solution. From the definition of E we also obtain convergence in the measures $\|\dot{\mathbf{u}}\|, \|\nabla \mathbf{u}\|, \|\mathbf{u}\|, \|p\|$ and $\|q\|$.

7.3 Exercises

Exercise 7.1. Let (u_i, p, q, θ) be a solution to the displacement boundary-initial value problem $\mathscr{P}$ for equations (2.34), i.e. with the boundary conditions

$$\begin{aligned} u_i(\mathbf{x},t) &= u_i^B(\mathbf{x},t), \quad p(\mathbf{x},t) = p^B(\mathbf{x},t), \\ q(\mathbf{x},t) &= q^B(\mathbf{x},t), \quad \theta(\mathbf{x},t) = \theta^B(\mathbf{x},t), \end{aligned}$$

on $\Gamma \times (0,T)$, and the initial conditions

$$\begin{aligned} &u_i(\mathbf{x},0) = v_i(\mathbf{x}), \quad \dot{u}_i(\mathbf{x},0) = w_i(\mathbf{x}), \quad p(\mathbf{x},0) = P(\mathbf{x}), \\ &q(\mathbf{x},0) = Q(\mathbf{x}), \quad \theta(\mathbf{x},0) = R(\mathbf{x}), \qquad \mathbf{x} \in \Omega. \end{aligned}$$

Suppose a_{ijkh} satisfy the symmetries (7.4) with the other coefficients being as in section 7.1 and with a_{ij} being symmetric. Use a logarithmic convexity method along the lines of sections 7.1, 7.1.1 and 7.1.2, *mutatis mutandis*, to demonstrate that the solution to $\mathscr{P}$ depends Hölder continuously upon the initial data.

Exercise 7.2. Let (u_i, p, q, s, θ) be a solution to the displacement boundary-initial value problem $\mathscr{P}$ for equations (2.47), i.e. with the boundary conditions

$$\begin{aligned} &u_i(\mathbf{x},t) = u_i^B(\mathbf{x},t), \quad p(\mathbf{x},t) = p^B(\mathbf{x},t), \quad q(\mathbf{x},t) = q^B(\mathbf{x},t), \\ &s(\mathbf{x},t) = s^B(\mathbf{x},t), \quad \theta(\mathbf{x},t) = \theta^B(\mathbf{x},t), \end{aligned}$$

on $\Gamma \times (0,T)$, and the initial conditions

$$\begin{aligned} &u_i(\mathbf{x},0) = v_i(\mathbf{x}), \quad \dot{u}_i(\mathbf{x},0) = w_i(\mathbf{x}), \quad p(\mathbf{x},0) = P(\mathbf{x}), \\ &q(\mathbf{x},0) = Q(\mathbf{x}), \quad s(\mathbf{x},0) = S(\mathbf{x}), \quad \theta(\mathbf{x},0) = R(\mathbf{x}), \qquad \mathbf{x} \in \Omega. \end{aligned}$$

Suppose a_{ijkh} satisfy the symmetries (7.4) with the other coefficients being as in section 7.1 and with a_{ij} being symmetric. Use a logarithmic convexity method along the lines of sections 7.1, 7.1.1 and 7.1.2, *mutatis mutandis*, to demonstrate that the solution to $\mathscr{P}$ depends Hölder continuously upon the initial data.

Exercise 7.3. Consider the boundary-initial value problem $\mathscr{V}$ for the isothermal version of the two voids equations (3.69), i.e. (u_i, ν, ω) satisfies

$$\begin{aligned} \rho \ddot{u}_i &= (a_{ijkh}u_{k,h})_{,j} + (b_{ij}\nu)_{,j} + (d_{ij}\omega)_{,j}\,, \\ \kappa_1 \ddot{\nu} &= (\alpha_{ij}\nu_{,j})_{,i} + (\beta_{ij}\omega_{,j})_{,i} - \alpha_1 \nu - \alpha_3 \omega - b_{ij}u_{i,j}\,, \\ \kappa_2 \ddot{\omega} &= (\gamma_{ij}\omega_{,j})_{,i} + (\beta_{ij}\nu_{,j})_{,i} - \alpha_2 \omega - \alpha_3 \nu - d_{ij}u_{i,j}\,, \end{aligned} \tag{7.90}$$

together with the boundary conditions

$$\begin{aligned} &u_i(\mathbf{x},t) = u_i^B(\mathbf{x},t), \quad \nu(\mathbf{x},t) = \nu^B(\mathbf{x},t), \\ &\omega(\mathbf{x},t) = \omega^B(\mathbf{x},t), \end{aligned} \tag{7.91}$$

on $\Gamma \times (0,T)$, and the initial conditions

$$\begin{aligned} u_i(\mathbf{x},0) &= v_i(\mathbf{x}), \quad \dot{u}_i(\mathbf{x},0) = w_i(\mathbf{x}), \\ \nu(\mathbf{x},0) &= \phi_1(\mathbf{x}), \quad \dot{\nu}(\mathbf{x},0) = \phi_2(\mathbf{x}), \\ \omega(\mathbf{x},0) &= \psi_1(\mathbf{x}), \quad \dot{\omega}(\mathbf{x},0) = \psi_2(\mathbf{x}), \end{aligned} \tag{7.92}$$

$\mathbf{x} \in \Omega$, where the functions $u_i^B, \dots, \psi_2$ are prescribed. Use a logarithmic convexity method along the lines of sections 1.4.4 - 1.4.6 to demonstrate Hölder continuous dependence upon the initial data when a_{ijkh} are required to only satisfy the symmetries (7.4).

Exercise 7.4. Let (u_i, ν, ω) satisfy the boundary-initial value problem $\mathscr{V}$ of exercise 7.3 but where the boundary conditions are zero ones, i.e.

$$u_i(\mathbf{x},t) = 0, \quad \nu(\mathbf{x},t) = 0, \quad \omega(\mathbf{x},t) = 0, \qquad \text{on } \Gamma \times \{t > 0\},$$

with the solution still satisfying the initial conditions (7.92).

Suppose a_{ijkh} satisfy the symmetries (7.4) but are not sign - definite. Define $F(t)$ by

$$F(t) = <\rho u_i u_i> + <\kappa_1 \nu^2> + <\kappa_2 \omega^2> + \beta(t+\tau)^2,$$

where β and τ are positive constants chosen in a manner analogous to that of exercise 1.14. Show that the solution to $\mathscr{V}$ in this case will grow exponentially as in (1.104) if $E(0) < 0$, or if $E(0) = 0$ and $<\rho v_i w_i> + <\kappa_1 \phi_1 \phi_2> + <\kappa_2 \psi_1 \psi_2> > 0$, where $E(t)$ is the energy function associated to $\mathscr{V}$.

Chapter 8
Waves in Double Porosity Elasticity

8.1 Rayleigh, Love, and Harmonic Waves

Rayleigh waves and Love waves have long been used as an important tool in geophysical problems employing linear elasticity theory. Rayleigh waves are studied using the Berryman & Wang [15] double porosity theory presented in section 2.2, by Dai *et al.* [59]. Dai *et al.* [59] argue that certain cross coefficients linking the macro pores and micro pores may be neglected in wave propagation problems and they use instead of equations (2.7) the system

$$\begin{aligned} &\rho_{11}\ddot{u}_i + \rho_{12}\ddot{v}_i + \rho_{13}\ddot{w}_i + (b_{12}+b_{13})\dot{u}_i - b_{12}\dot{v}_i - b_{13}\dot{w}_i \\ &\qquad = \Big(K_u + \frac{\mu}{3}\Big)u_{j,ji} + \mu\Delta u_i + K_u(B_1\zeta^1_{,i} + B_2\zeta^2_{,i}), \\ &\rho_{12}\ddot{u}_i + \rho_{22}\ddot{v}_i - b_{12}\dot{u}_i + (b_{12}+b_{23})\dot{v}_i = -(1-v_2)\phi^1 p_{,i}, \\ &\rho_{13}\ddot{u}_i + \rho_{33}\ddot{w}_i - b_{13}\dot{u}_i + (b_{13}+b_{23})\dot{w}_i = -v_2\phi^2 q_{,i}. \end{aligned} \tag{8.1}$$

They write the solution (u_i, v_i, w_i) in the form

$$\mathbf{u} = \nabla\phi_s + \nabla\psi_s, \quad \mathbf{v} = \nabla\phi_1 + \nabla\psi_1, \quad \mathbf{w} = \nabla\phi_2 + \nabla\psi_2,$$

where $\phi_s, \phi_1, \phi_2, \psi_s, \psi_1$ and ψ_2 are potential functions.

For a Rayleigh wave Dai *et al.* [59] take the elastic body to be one with a double porosity strcture occupying the half space $\mathbb{R}^2 \times \{z > 0\}$ with the $z-$axis pointing downward. As they are seeking a Rayleigh wave they look for functions $\phi_s, \ldots, \psi_2$ of form

$$\phi_s = A_s e^{-\lambda z} e^{i(kx-\omega t)}, \phi_1 = A_f e^{-\lambda z} e^{i(kx-\omega t)}, \phi_2 = A_g e^{-\lambda z} e^{i(kx-\omega t)},$$

and

$$\psi_s = \mathbf{B}_s e^{-sz} e^{i(kx-\omega t)}, \quad \psi_1 = \mathbf{B}_f e^{-sz} e^{i(kx-\omega t)}, \quad \psi_2 = \mathbf{B}_g e^{-sz} e^{i(kx-\omega t)},$$

B. Straughan, *Mathematical Aspects of Multi–Porosity Continua*, Advances in Mechanics and Mathematics 38, https://doi.org/10.1007/978-3-319-70172-1_8

where k is a complex wavenumber and λ and s ensure decay into the ground, i.e. as $z \to \infty$.

Dai *et al.* [59] present a comprehensive analysis of Rayleigh waves, varying material properties appropriate to a real rock, namely Berea sandstone.

It would be interesting to analyse Rayleigh waves on an exponentially graded double porosity elastic half space, cf. the analysis for a single void distribution in Chirita [38].

Love waves are analysed by Dai & Kuang [58] who again employ the double porosity elasticity theory of Berryman & Wang [15]. They argue that Love waves involve only the horizontal component in an oscillatory motion and so they may use the Berryman & Wang [15] equations given in (2.7) but neglecting the x and z components. Thus, they work with the system of equations for (u_2, v_2, w_2) of the form

$$\begin{aligned}
&\rho_{11}\ddot{u}_2 + \rho_{12}\ddot{v}_2 + \rho_{13}\ddot{w}_2 + (b_{12}+b_{13})\dot{u}_2 - b_{12}\dot{v}_2 - b_{13}\dot{w}_2 = \sigma_{2j,j}\,,\\
&\rho_{12}\ddot{u}_2 + \rho_{22}\ddot{v}_2 + \rho_{23}\ddot{w}_2 - b_{12}\dot{u}_2 + (b_{12}+b_{23})\dot{v}_2 - b_{23}\dot{w}_2 = -(1-v_2)\phi^1 p_{,2}\,, \qquad (8.2)\\
&\rho_{13}\ddot{u}_2 + \rho_{23}\ddot{v}_2 + \rho_{33}\ddot{w}_2 - b_{13}\dot{u}_2 - b_{23}\dot{v}_2 + (b_{13}+b_{23})\dot{w}_2 = -v_2\phi^2 q_{,2}\,.
\end{aligned}$$

Dai & Kuang [58] further neglect cross terms coupling the macro and micro pores and they seek solutions of form

$$u_2 = f(z)\,e^{i(kx-\omega t)}, \quad v_2 = g(z)\,e^{i(kx-\omega t)}, \quad w_2 = h(z)\,e^{i(kx-\omega t)}.$$

The functions f, g and h are calculated from the governing differential equations and appropriate boundary conditions are employed. Detailed numerical solutions are presented for three cases. These are when an elasic layer is overlying a double porosity half space, when a double porosity layer is overlying an elastic half space, and when a double porosity layer is overlying a double porosity half space.

Harmonic waves are analysed for the Svanadze equations (2.35) - (2.37) and (2.38) - (2.40) by Svanadze [212] and by Ciarletta *et al.* [51].

Svanadze [212] analyses plane harmonic waves in an isotropic linearly elastic body which has a double porosity structure. His analysis is based on equations (2.35) - (2.37), although he chooses to ignore the elastic acceleration term. Thus, Svanadze [212] employs the equations

$$0 = \mu\Delta u_i + (\lambda+\mu)\frac{\partial}{\partial x_i}\Big(\frac{\partial u_j}{\partial x_j}\Big) - \hat{\beta}\frac{\partial p}{\partial x_i} - \hat{\gamma}\frac{\partial q}{\partial x_i}\,, \tag{8.3}$$

and

$$\alpha\dot{p} = k\Delta p - \gamma(p-q) - \hat{\beta}\frac{\partial \dot{u}_i}{\partial x_i}\,, \tag{8.4}$$

coupled with

$$\beta\dot{q} = m\Delta q + \gamma(p-q) - \hat{\gamma}\frac{\partial \dot{u}_i}{\partial x_i}\,. \tag{8.5}$$

Since he is dealing with harmonic waves Svanadze [212] chooses solutions which have form

$$\begin{aligned} u_m(\mathbf{x},t) &= A_m \exp\{i(\Lambda x-\omega t)\}, \qquad m=1,2,3, \\ p(\mathbf{x},t) &= P\exp\{i(\Lambda x-\omega t)\}, \\ q(\mathbf{x},t) &= Q\exp\{i(\Lambda x-\omega t)\}. \end{aligned} \tag{8.6}$$

In equations (8.6) the parameters A_1, A_2, A_3, P and Q are constants, $\omega > 0$ is the frequency, and Λ is a complex number. After substituting these expressions into the equations (8.3) - (8.5) Svanadze [212] deduces there are five equations relating ω and Λ. He uses these equations to derive dispersion equations for longitudinal and for transverse waves. Both longitudinal and transverse waves are analysed in detail.

A further study of plane harmonic waves in an isotropic linearly elastic body with double porosity is by Ciarletta *et al.* [51]. The analysis of Ciarletta *et al.* [51] employs the full equations (2.38) - (2.40) and so they include the cross inertia pressure coefficients as well as the acceleration in the momentum equation. Again, a solution is sought of a similar form to (8.6) and dispersion equations are derived for longitudinal and for transverse waves. The dispersion relations are investigated in detail and a comparison is made with harmonic waves in classical elasticity. In this way Ciarletta *et al.* [51] are able to ascertain the effects of the macro and micro porosities.

8.2 Acceleration Waves

Our aim in this section is to produce a fully nonlinear theory generalizing the Svanadze model of elasticity with double porosity in section 2.3, and then investigate an acceleration wave analysis of the new equations. Gentile & Straughan [90] have already produced an acceleration wave analysis for such a nonlinear elasticity theory with double porosity by generalizing equations (2.25), (2.26) and (2.27). In this work we adopt a different approach and produce a generalization of the work of Biot [19].

Biot [19] developed a theory for a nonlinear elastic body based on a strain energy function, but he allowed for pores in the elastic material. He took the momentum equation to hold in the equilibrium case, i.e. he neglected the acceleration term in the momentum equation. He accounted for the pore structure by including a nonlinear parabolic equation for the pressure distribution in the pores, where the time dependence of the pressure field is explicitly accounted for. As Biot [19] writes, ... "*The mechanics of porous media is thus brought to the same level of development of the classical theory of finite deformations in elasticity*". Biot's [19] work is not based on a continuum thermodynamic approach employing an entropy inequality. While Biot [19] uses an equilibrium version of the momentum equation and treats the isothermal situation he writes, ... "*In order to restrict the length of the paper, the theory is presented in the context of quasi-static and isothermal deformations*".

8.2.1 Nonlinear Elasticity with Double Porosity

To develop a nonlinear theory for an elastic body with a double porosity, namely a macro porosity and a micro porosity, we use the notation of sections 3.1.1, 3.1.2, 3.1.3 and 3.2. Thus, we begin with the equation for the balance of linear momentum as

$$\rho_0 \ddot{x}_i = \frac{\partial \Pi_{Ai}}{\partial X_A} + \rho_0 f_i \,, \tag{8.7}$$

where Π_{Ai} is the Piola-Kirchoff stress tensor given by

$$\Pi_{Ai} = \rho_0 \frac{\partial \psi}{\partial F_{iA}} \,, \tag{8.8}$$

where ψ is the Helmholtz free energy function. The points x_i in the current configuration and X_A in the reference configuration are as in section 3.1.1 and the deformation gradient tensor is defined as in (3.4), or, see also (3.5).

Let p and q denote the pressures in the macro pores and micro pores, respectively. Then we define the constitutive list χ by

$$\chi = \{F_{iA}, p, q, X_K\} \,. \tag{8.9}$$

In this work the Helmholtz free energy function is taken to depend on the variables in the list χ, i.e.

$$\psi = \psi(\chi) \equiv \psi\{F_{iA}, p, q, X_K\} \,. \tag{8.10}$$

We now require equations for the pressure fields themselves and we follow the approach in Biot [19] leading to his equation (3.26) although we additionally have to account for interactions between the pore pressures. Thus, for functions m and n which depend on the variables in the constitutive list χ and for fluxes J_A and K_A we have governing conservation equations

$$\frac{\partial m}{\partial t} = \frac{\partial J_A}{\partial X_A} + I \,, \tag{8.11}$$

and

$$\frac{\partial n}{\partial t} = \frac{\partial K_A}{\partial X_A} - I \,, \tag{8.12}$$

where $I(p,q)$ represents the interactions between the pressures in the macro and the micro pores. In the linear case $I = -\gamma(p-q)$ as seen in equation (2.26). The fluxes J_A and K_A are taken to be functions of the variables

$$\begin{aligned} J_A &= J_A(F_{iR}, p, q, p_{,R}, q_{,R}, X_K), \\ K_A &= K_A(F_{iR}, p, q, p_{,R}, q_{,R}, X_K). \end{aligned} \tag{8.13}$$

To be specific note that

$$m = m(F_{iR}, p, q, X_K),$$
$$n = n(F_{iR}, p, q, X_K). \tag{8.14}$$

The governing equations for a nonlinear elastic body with a double porosity structure are then (8.7), (8.11) and (8.12). For an acceleration wave analysis it is sufficient to take $f_i = 0$ in equation (8.7).

We define an acceleration wave $\mathscr{S}$ for equations (8.7), (8.11) and (8.12) to be a singular surface $\mathscr{S}$ across which $\ddot{x}_i, \dot{x}_{i,A}, x_{i,AB}, \ddot{p}, \dot{p}_{,A}, p_{,AB}, \ddot{q}, \dot{q}_{,A}$ and $q_{,AB}$ and their higher derivatives suffer a finite discontinuity, but $x_i, p, q \in C^1(\mathbb{R}^3 \times [0,T])$, $[0,T]$ being the time interval.

To develop a fully nonlinear acceleration wave analysis we need some compatibility relations. We begin by recalling the definition of the jump of a function f, denoted by $[f]$, as given by (4.6).

General compatibility relations for a function $\psi(\mathbf{X},t)$ are needed across $\mathscr{S}$. These are given in detail in Truesdell & Toupin [221] or in Chen [34]. We simply quote those we need. If ψ is continuous in $\mathbb{R}^3$ but its derivative is discontinuous across $\mathscr{S}$ then

$$[\psi_{,A}] = N_A B, \qquad \text{where } B = [N^R \psi_{,R}]. \tag{8.15}$$

When $\psi \in C^1(\mathbb{R}^3)$ then

$$[\psi_{,AB}] = N_A N_B C, \qquad \text{where } C = [N^R N^S \psi_{,RS}]. \tag{8.16}$$

In (8.15) and (8.16), N_A refers to the unit normal to $\mathscr{S}$, but referred back to the reference configuration. Relations (8.15) and (8.16) are derived from Chen [34], equations (4.13), (4.14). The Hadamard formula in three dimensions is, cf. Chen [34] (4.15),

$$\frac{\delta}{\delta t}[\psi] = [\dot{\psi}] + U_N B \tag{8.17}$$

where $\dot{\psi} = \partial \psi / \partial t|_{\mathbf{X}}$, U_N is the speed at the point on $\mathscr{S}$ with unit normal N_A and B is defined in (8.15).

Since $u_i \in C^1(\mathbb{R}^3)$ we find using (8.17)

$$0 = \frac{\delta}{\delta t}[\dot{u}_i] = [\ddot{u}_i] + U_N [N^A \dot{u}_{i,A}] \tag{8.18}$$

$$0 = \frac{\delta}{\delta t}[u_{i,A}] = [\dot{u}_{i,A}] + U_N [N^B u_{i,AB}]. \tag{8.19}$$

Whence from (8.18) and (8.19) we derive

$$[\ddot{u}_i] = -U_N [N^A \dot{u}_{i,A}] \tag{8.20}$$

$$[N^A \dot{u}_{i,A}] = -U_N [N^A N^B u_{i,AB}]. \tag{8.21}$$

Hence, combining (8.20) and (8.21) one finds

$$[\ddot{u}_i] = U_N^2 [N^A N^B u_{i,AB}]. \tag{8.22}$$

By repeated use of (8.15) we find

$$\begin{aligned}[u_{r,AB}] &= N_A[N^R u_{r,BR}] \\ &= N_A N_B [N^R N^S u_{r,RS}] \\ &= \frac{N_A N_B}{U_N^2}[\ddot{u}_r] \end{aligned} \tag{8.23}$$

where in the last line we have employed (8.22).

Next, employ the chain rule on equations (8.7), (8.11) and (8.12), using the constitutive theory (8.10), (8.13) and (8.14). In this way one derives

$$\ddot{x}_i = \frac{\partial^2 \psi}{\partial F_{jB} \partial F_{iA}} x_{j,AB} + \frac{\partial^2 \psi}{\partial p \partial F_{iA}} p_{,A} + \frac{\partial^2 \psi}{\partial q \partial F_{iA}} q_{,A} + \frac{\partial}{\partial X_A}\Big(\frac{\partial \psi}{\partial F_{iA}}\Big)\Big|_{\mathbf{F},\nabla p,\nabla q}, \tag{8.24}$$

and

$$\begin{aligned} \frac{\partial m}{\partial F_{iA}} \dot{x}_{i,A} + \frac{\partial m}{\partial p} \dot{p} + \frac{\partial m}{\partial q} \dot{q} = & \frac{\partial J_A}{\partial F_{jB}} x_{j,BA} \\ & + \frac{\partial J_A}{\partial p} p_{,A} + \frac{\partial J_A}{\partial q} q_{,A} + \frac{\partial J_A}{\partial p_{,R}} p_{,RA} \\ & + \frac{\partial J_A}{\partial q_{,R}} q_{,RA} + \frac{\partial J_A}{\partial X_A}\Big|_{\mathbf{F},p,q,\nabla p,\nabla q} + I, \end{aligned} \tag{8.25}$$

and

$$\begin{aligned} \frac{\partial n}{\partial F_{iA}} \dot{x}_{i,A} + \frac{\partial n}{\partial p} \dot{p} + \frac{\partial n}{\partial q} \dot{q} = & \frac{\partial K_A}{\partial F_{jB}} x_{j,BA} \\ & + \frac{\partial K_A}{\partial p} p_{,A} + \frac{\partial K_A}{\partial q} q_{,A} + \frac{\partial K_A}{\partial p_{,R}} p_{,RA} \\ & + \frac{\partial K_A}{\partial q_{,R}} q_{,RA} + \frac{\partial K_A}{\partial X_A}\Big|_{\mathbf{F},p,q,\nabla p,\nabla q} - I, \end{aligned} \tag{8.26}$$

where $\nabla p \equiv \partial p / \partial X_A$ and where the terms evaluated at constant $\mathbf{F}, p, q, \nabla p, \nabla q$ are due to the nonhomogeneous effect of the inclusion of X_K in (8.10), (8.13) and (8.14), cf. the presentation in thermoelasticity, equations (3.35) and (3.36).

Now take the jumps of equations (8.24) - (8.26) to find, recalling the definition of an acceleration wave,

$$[\ddot{x}_i] = \frac{\partial^2 \psi}{\partial F_{jB} \partial F_{iA}} [x_{j,AB}], \tag{8.27}$$

and

$$\frac{\partial m}{\partial F_{iA}} [\dot{x}_{i,A}] = \frac{\partial J_A}{\partial F_{jB}} [x_{j,BA}] + \frac{\partial J_A}{\partial p_{,R}} [p_{,RA}] + \frac{\partial J_A}{\partial q_{,R}} [q_{,RA}], \tag{8.28}$$

and

$$\frac{\partial n}{\partial F_{iA}} [\dot{x}_{i,A}] = \frac{\partial K_A}{\partial F_{jB}} [x_{j,BA}] + \frac{\partial K_A}{\partial p_{,R}} [p_{,RA}] + \frac{\partial K_A}{\partial q_{,R}} [q_{,RA}]. \tag{8.29}$$

Next, employ relation (8.23) to see that

$$[x_{j,BA}] = \frac{N_A N_B}{U_N^2} [\ddot{x}_j].$$

We employ this relation in (8.27). Define the amplitudes a_i, b and c by

$$a_i(t) = [\ddot{x}_i], \qquad b(t) = [\ddot{p}], \qquad c(t) = [\ddot{q}].$$

Then (8.27) leads to

$$U_N^2 a_i = \frac{\partial^2 \psi}{\partial F_{jB} \partial F_{iA}} N_A N_B a_j .$$

If we recall (8.8) this can be rewritten in a more familiar form as

$$\rho_0 U_N^2 a_i = Q_{ij} a_j \tag{8.30}$$

where Q_{ij} is the acoustic tensor given by

$$Q_{ij} = N_B N_A \frac{\partial \Pi_{Ai}}{\partial F_{jB}} . \tag{8.31}$$

One now uses the relation

$$N_A = F_{iA} n_i \frac{|\nabla_{\mathbf{x}} s|}{|\nabla_{\mathbf{X}} \mathscr{S}|} \tag{8.32}$$

see Truesdell & Toupin [221], equation (182.8), where s and n_i and $\mathscr{S}$ and N_A are the surface and unit normal to the surface in the current and reference coordinates, respectively. In this way one rewrites $Q_{ij}(N_A, U_N)$ as a function $Q_{ij}(n_r, U_N)$ and then from (8.30) one deduces that an acceleration wave may propagate provided a_i is an eigenvector of Q_{ij}, cf. Truesdell & Noll [220], p. 271, for the case of classical nonlinear elasticity. The question of existence of an acceleration wave and which particular direction it will take is discussed at length in Truesdell & Noll [220], section 71, and in Truesdell [219] and in Chen [34], pp. 316–322. The arguments given there may be applied to equation (8.30). In this way we may assert that a longitudinal wave will propagate if $Q_{ij}(\mathbf{n})$ is positive definite, i.e. if

$$n_i Q_{ij} n_j > 0$$

for all unit vectors $\mathbf{n}$, or, if the strong ellipticity condition holds, i.e.

$$\lambda_i Q_{ij}(\mathbf{n}) \lambda_j > 0$$

for all unit vectors n_i and λ_i. In general, once one knows the direction of the acceleration wave, say $a_i = a(t)\lambda_i$, for a unit vector λ_i, then taking the inner product of (8.30) with λ_i leads to the wavespeed U_N (of a left and a right moving acceleration wave), namely

$$U_N^2 = \frac{Q_{ij} \lambda_i \lambda_j}{\rho_0} .$$

The remaining equations (8.28) and (8.29) lead to equations for the other amplitudes b and c. To see this we use (8.15), (8.16), (8.20), (8.21) and (8.22) and their equivalents for p and q to rewrite (8.28) and (8.29) in the forms

$$\frac{\partial J_A}{\partial p_{,R}} N_A N_R b + \frac{\partial J_A}{\partial q_{,R}} N_A N_R c = -\left(U_N N_A \frac{\partial m}{\partial F_{iA}} + \frac{\partial J_A}{\partial F_{iB}} N_A N_B \right) a_i \tag{8.33}$$

and

$$\frac{\partial K_A}{\partial p_{,R}} N_A N_R b + \frac{\partial K_A}{\partial q_{,R}} N_A N_R c = -\left(U_N N_A \frac{\partial n}{\partial F_{iA}} + \frac{\partial K_A}{\partial F_{iB}} N_A N_B \right) a_i . \tag{8.34}$$

Once a_i is known these are just a pair of simultaneous equations for the pressure amplitudes b and c.

8.2.2 Amplitude Equation

To derive the amplitude equation we restrict attention to the one-dimensional space situation. One may proceed in the three-dimensional case but the differential geometry involved may obscure the basic physics, cf. Lindsay & Straughan [152]. Hence, we rewrite equations (8.7) - (8.12) for one spatial variable denoted by x in the current configuration and X in the reference configuration. One may employ the displacement $u = x - X$ and then the governing equations are

$$\ddot{u} = \frac{\partial \psi_F}{\partial X}, \tag{8.35}$$

and

$$\dot{m} = \frac{\partial J}{\partial X} + I, \tag{8.36}$$

together with

$$\dot{n} = \frac{\partial K}{\partial X} - I. \tag{8.37}$$

Here $F = u_X$,

$$\psi = \psi(F, p, q),$$

and

$$J = J(F, p, q, p_X, q_X), \qquad K = K(F, p, q, p_X, q_X), \qquad I = I(p, q).$$

The jump equations from (8.35) - (8.37) are

$$\begin{aligned} [\ddot{u}] &= \psi_{FF}[u_{XX}], \\ m_F[\dot{u}_X] &= J_F[u_{XX}] + J_{p_X}[p_{XX}] + J_{q_X}[q_{XX}], \\ n_F[\dot{u}_X] &= K_F[u_{XX}] + K_{p_X}[p_{XX}] + K_{q_X}[q_{XX}]. \end{aligned} \tag{8.38}$$

Let us denote the wavespeed in one-dimension by V then the Hadamard relation in this case is

$$\frac{\delta}{\delta t}[f] = [f_t] + V[f_X] \equiv [\dot{f}] + V[f_X]\,. \tag{8.39}$$

We now restrict attention to a wave moving into a state of equilibrium in the sense that $u_X^+ =$constant, $p^+ =$constant, $q^+ =$constant.

By employing the Hadamard relation one finds

$$[\ddot{u}] = -V[\dot{u}_X] = V^2[u_{XX}]\,,$$

and then from $(8.38)_1$ one obtains

$$(V^2 - \psi_{FF})[u_{XX}] = 0\,,$$

whence

$$V^2 = \psi_{FF}\,, \tag{8.40}$$

and V is here constant since the wave moves into an equilibrium region. Introduce the wave amplitudes a, b and c by

$$a(t) = [\ddot{u}], \quad b(t) = [\ddot{p}], \quad c(t) = [\ddot{q}]\,,$$

then employing the Hadamard relation equations $(8.38)_2$ and $(8.38)_3$ may be written as

$$\begin{aligned} J_{pX}[p_{XX}] + J_{qX}[q_{XX}] &= -\Big(\frac{J_F}{V^2} + \frac{m_F}{V}\Big)a, \\ K_{pX}[p_{XX}] + K_{qX}[q_{XX}] &= -\Big(\frac{K_F}{V^2} + \frac{n_F}{V}\Big)a. \end{aligned} \tag{8.41}$$

These equations allow us to show

$$[p_{XX}] = -\frac{\alpha_1}{V^2}a \qquad \text{and} \qquad [q_{XX}] = -\frac{\alpha_2}{V^2}a\,, \tag{8.42}$$

where

$$\begin{aligned} \alpha_1 &= \frac{1}{J}\{K_{qX}(J_F + Vm_F) - J_{qX}(K_F + Vn_F)\}\,, \\ \alpha_2 &= \frac{1}{J}\{-K_{pX}(J_F + Vm_F) + J_{pX}(K_F + Vn_F)\}\,. \end{aligned}$$

To proceed with the amplitude calculation we differentiate (8.35) with respect to X and take the jumps of the resulting equation recalling u_X^+, p^+, q^+ are constants to obtain

$$[\ddot{u}_X] = \psi_{FF}[u_{XXX}] + \psi_{FFF}[u_{XX}^2] + \psi_{F_p}[p_{XX}] + \psi_{F_q}[q_{XX}]\,. \tag{8.43}$$

We now employ (8.42) to remove the $[p_{XX}]$ and $[q_{XX}]$ terms. The $[u_{XX}^2]$ term is handled using the jump of a product

$$[gh] = g^+[h] + h^+[g] + [g][h]\,. \tag{8.44}$$

We also use the Hadamard relation to show

$$\begin{aligned}-[\ddot{u}_X]+\psi_{FF}[u_{XXX}] &= \frac{2}{V}\frac{\delta a}{\delta t}+(\psi_{FF}-V^2)[u_{XXX}]\\ &= \frac{2}{V}\frac{\delta a}{\delta t}.\end{aligned}$$

In this way from (8.43) we arrive at the amplitude equation

$$\frac{\delta a}{\delta t}+\frac{\psi_{FFF}}{2V^3}a^2-\alpha_3 a=0, \tag{8.45}$$

where

$$\alpha_3=\frac{\alpha_1}{2V}\psi_{F_p}+\frac{\alpha_2}{2V}\psi_{F_q}.$$

It is interesting to compare (8.45) with the equivalent equation for nonlinear elasticity, i.e. when there are no pores present. The equation then is the same as (8.45) but with $\alpha_3 = 0$, see e g. Straughan [202], equation (7.21). Equation (8.45) clearly displays a damping effect included in the acceleration wave behaviour due to the macro and micro porosities. According to the current theory the wavespeed in the purely elastic case and the double porosity case are the same, but there is a strong effect on the amplitude due to the presence of both sets of pores.

The coefficients of (8.45) are constant and we let $k = \psi_{FFF}/2V^3$. The solution to (8.45) is

$$a(t)=\frac{a(0)}{e^{-\alpha_3 t}+(ka(0)/\alpha_3)(1-e^{-\alpha_3 t})}. \tag{8.46}$$

If $a(0) < 0$ then the wave amplitude will blow-up in a finite time

$$T=\frac{1}{\alpha_3}\log\left(\frac{ka(0)-\alpha_3}{ka(0)}\right).$$

This blow-up time may be contrasted with the analogous one in the purely elastic case, where cf. Straughan [202], p. 304, $T_E = -1/ka(0)$.

The wave amplitude $a(t)$ is given explicitly by (8.46) and we point out that the pressure amplitudes b and c are then determined from (8.42) using the relations $b = V^2[p_{XX}]$ and $c = V^2[q_{XX}]$.

8.3 Exercises

Exercise 8.1. The equations of nonlinear elasticity are, cf. (3.7), with zero body force,

$$\rho_0\ddot{x}_i=\frac{\partial \Pi_{iA}}{\partial X_A},$$

where $\Pi_{Ai} = \rho_0 \partial\psi/\partial F_{iA}$ is the Piola-Kirchoff stress tensor. The Helmholtz free energy function is such that $\psi = \psi(F_{iA})$. Find the wavespeed equation for an acceleration wave and analyse possible wave motion.

Exercise 8.2. Find the general amplitude equation in one-dimensional nonlinear elasticity when

$$\psi = \frac{\alpha}{2}u_X^2 + \frac{\beta}{3}u_X^3 + \frac{\gamma}{4}u_X^4,$$

α, β, γ constants. Show that for a wave moving into a region where $x = \lambda X$ then the wavespeed arises from

$$V^2 = \frac{1}{\rho_0}\{\alpha + 2\beta(\lambda - 1) + 3\gamma(\lambda - 1)^2\}.$$

Show that the amplitude equation is

$$\frac{\delta a}{\delta t} + ka^2 = 0,$$

where

$$k = \frac{\{\beta + 3\gamma(\lambda - 1)\}\sqrt{\rho_0}}{\{\alpha + 2\beta(\lambda - 1) + 3\gamma(\lambda - 1)^2\}^{3/2}}$$

Deduce that

$$a(t) = \frac{a(0)}{1 + a(0)kt},$$

and $a(t)$ blows-up in finite time $T = -1/a(0)k$ provided $a(0)k < 0$, cf. Straughan [202], p. 304.

Exercise 8.3. Consider a Biot nonlinear model for an elastic body with single pores given by equations (8.7) and (8.11), so with zero body force one has

$$\begin{aligned} \rho_0 \ddot{x}_i &= \frac{\partial \Pi_{iA}}{\partial X_A}, \\ \frac{\partial m}{\partial t} &= \frac{\partial J_A}{\partial X_A}. \end{aligned} \tag{8.47}$$

The Piola-Kirchoff stress tensor is given by (8.8) and the functions ψ and J_A satisfy the constitutive theory

$$\psi = \psi(F_{iR}, p), \qquad J_A = J_A(F_{iR}, p, p_{,R}).$$

Define an acceleration wave for equations (8.47) and derive the jump equations which arise from (8.47). Derive the wavespeed equations and discuss the propagation conditions.

Exercise 8.4. For the one-dimensional version of exercise 8.3 derive the amplitude equation for an acceleration wave moving into an equilibrium region for which u_X^+ =constant, p^+ =constant, and solve it.

Discuss the conditions under which the amplitude may exhibit finite time blow-up.

Exercise 8.5. By using arguments similar to those of section 3.2.2 derive the linear equations of Svanadze [212, 214], namely (2.28) - (2.30), or (2.25) - (2.27), from equations (8.24) - (8.26). You should need the linear constitutive equations

$$
\begin{aligned}
t_{ij} &= a_{ijkh}u_{k,h} - \beta_{ij}p - \gamma_{ij}q\,, \\
m &= \alpha_1 p + \alpha_2 q + \beta_{ij}e_{ij}\,, \\
n &= \beta_1 p + \beta_2 q + \gamma_{ij}e_{ij}\,,
\end{aligned}
$$

and an argument like that leading to (3.27).

Chapter 9
Acceleration Waves in Double Voids

9.1 Isothermal Acceleration Waves

In this chapter we develop a fully nonlinear acceleration wave analysis for the double voids elasticity theory of Iesan & Quintanilla [115]. We commence with the isothermal case and so the starting point is equations (3.55), (3.56) and (3.57). For an acceleration wave analysis it is sufficient to take the source terms to be zero and so the relevant equations are

$$\rho_0 \ddot{x}_i = \Pi_{Ai,A}\,, \tag{9.1}$$

with

$$\kappa_1 \ddot{\nu} = h_{A,A} + g\,, \tag{9.2}$$

and

$$\kappa_2 \ddot{\omega} = j_{A,A} + h\,. \tag{9.3}$$

The terms Π_{Ai}, h_A, j_A, g and h are given in terms of the Helmholtz free energy as

$$\begin{aligned} &\Pi_{Ai} = \rho_0 \frac{\partial \psi}{\partial F_{iA}}\,, \qquad h_A = \rho_0 \frac{\partial \psi}{\partial \nu_{,A}}\,, \qquad j_A = \rho_0 \frac{\partial \psi}{\partial \omega_{,A}}\,, \\ &g = -\rho_0 \frac{\partial \psi}{\partial \nu}\,, \qquad h = -\rho_0 \frac{\partial \psi}{\partial \omega}\,. \end{aligned} \tag{9.4}$$

The Helmholtz free energy ψ depends on the following variables

$$\psi = \psi(F_{iA}, \nu, \omega, \nu_{,A}, \omega_{,A}) \tag{9.5}$$

where $F_{iA} = x_{i,A}$. We also employ $u_{i,A}$ where $u_i = x_i - X_i$.

An acceleration wave for equations (9.1) - (9.3) is a singular surface $\mathscr{S}$ such that u_i, ν, ω are $C^1(\mathbb{R}^3 \times \{t > 0\})$ but $\ddot{u}_i, \dot{u}_{i,A}, u_{i,AB}$ $\ddot{\nu}, \dot{\nu}_{,A}, \nu_{,AB}$ $\ddot{\omega}, \dot{\omega}_{,A}$ and $\omega_{,AB}$ and their higher derivatives suffer a finite discontinuity across $\mathscr{S}$.

We define the wave amplitudes a_i, b and c by

$$a_i(t) = [\ddot{u}_i], \qquad b(t) = [\ddot{\nu}], \qquad c(t) = [\ddot{\omega}], \tag{9.6}$$

B. Straughan, *Mathematical Aspects of Multi–Porosity Continua*, Advances in Mechanics and Mathematics 38, https://doi.org/10.1007/978-3-319-70172-1_9

and then take the jumps of equations (9.1) - (9.3) after expanding terms using (9.4) and (9.5). After using the compatibility relations and (8.23) one may show that the jump equations lead to

$$\begin{aligned}(\rho_0 U_N^2 \delta_{ij} - Q_{ij})a_j &= J_i b + K_i c,\\ (\kappa_1 U_N^2 - Q_b)b &= J_i a_i + Q_r c,\\ (\kappa_2 U_N^2 - Q_c)c &= K_i a_i + Q_r b,\end{aligned} \tag{9.7}$$

where Q_{ij} is the acoustic tensor

$$Q_{ij} = \rho_0 N_A N_B \frac{\partial^2 \psi}{\partial F_{iA} \partial F_{jB}}. \tag{9.8}$$

The voids acoustic terms Q_b and Q_c are given by

$$\begin{aligned}Q_b &= N_A N_B \frac{\partial h_A}{\partial \nu_{,B}} = \rho_0 \frac{\partial^2 \psi}{\partial \nu_{,A} \partial \nu_{,B}} N_A N_B\,,\\ Q_c &= N_A N_B \frac{\partial j_A}{\partial \omega_{,B}} = \rho_0 \frac{\partial^2 \psi}{\partial \omega_{,A} \partial \omega_{,B}} N_A N_B\,,\end{aligned} \tag{9.9}$$

while the terms J_i, K_i and Q_r are

$$\begin{aligned}J_i &= N_A N_B \frac{\partial h_A}{\partial F_{iB}} = \rho_0 \frac{\partial^2 \psi}{\partial \nu_{,A} \partial F_{iB}} N_A N_B\,,\\ K_i &= N_A N_B \frac{\partial j_A}{\partial F_{iB}} = \rho_0 \frac{\partial^2 \psi}{\partial \omega_{,A} \partial F_{iB}} N_A N_B\,,\\ Q_r &= N_A N_B \frac{\partial h_A}{\partial \omega_{,B}} = N_A N_B \frac{\partial j_A}{\partial \nu_{,B}} = \rho_0 \frac{\partial^2 \psi}{\partial \omega_{,A} \partial \nu_{,B}} N_A N_B\,.\end{aligned} \tag{9.10}$$

Equations (9.7) are the propagation conditions which govern how and when an acceleration wave may propagate.

From this juncture there are various avenues to explore. For example, we could consider

(a) $a_i = a(t)n_i$, a longitudinal wave,
(b) $a_i = \hat{a}(t)s_i$, s_i is a tangential vector to $\mathscr{S}$, a transverse wave,
(c) body has a centre of symmetry, then $J_i = 0, K_i = 0$.

For example, in case (a) by taking the inner product of $(9.7)_1$ with n_i we may deduce the wavespeed equation as

$$\begin{aligned}&(\rho_0 U_N^2 - Q_{ij}n_i n_j)(\kappa_1 U_N^2 - Q_b)(\kappa_2 U_N^2 - Q_c) - (K_i n_i)^2(\kappa_1 U_N^2 - Q_b)\\ &\quad - (J_i n_i)^2(\kappa_2 U_N^2 - Q_c) - Q_r^2(\rho_0 U_N^2 - Q_{ij}n_i n_j)\\ &\quad - 2Q_r J_i n_i K_a n_a = 0.\end{aligned} \tag{9.11}$$

This is a cubic equation in U_N^2 which yields the possibility of three waves each moving in positive and negative n_i directions. These waves will arise due to an elastic wave and two other waves associated with the macro and the micro pores.

Let us return to situation (c) outlined above. The concept of a centre of symmetry involves the symmetry group of the material and whether orthogonal or proper orthogonal transformations are allowed. This topic is explained in Spencer [199], pp. 106 – 110, Ogden [168], pp. 180 – 183, 209 – 213, and Truesdell & Noll [220], pp. 76 – 81, 149 – 151, although for elastic materials not containing voids. Many materials are centrosymmetric and so it is useful to analyse this class of bodies. However, there are also many other types of material where effects such as chirality are necessary and then centrosymmetry is not imposed, see e.g. Lakes [149], Iesan & Quintanilla [115]. Such effects in multi-porosity elasticity are likely to lead to interesting new analyses.

For the centrosymmetric case $J_i = 0$ and $K_i = 0$ and then (9.7) reduce to

$$
\begin{aligned}
&(\rho_0 U_N^2 \delta_{ij} - Q_{ij})a_j = 0,\\
&(\kappa_1 U_N^2 - Q_b)b - Q_r c = 0,\\
&(\kappa_2 U_N^2 - Q_c)c - Q_r b = 0.
\end{aligned}
\tag{9.12}
$$

Equation $(9.12)_1$ is associated to the elastic wave whereas $(9.12)_{2,3}$ are asociated to coupled waves arising due to the presence of macro and micro voids. Equations $(9.12)_{2,3}$ yield for the wavespeed

$$
(U_N^2 - U_I^2)((U_N^2 - U_{II}^2) - K^2 = 0, \tag{9.13}
$$

where

$$
U_I^2 = \frac{Q_b}{\kappa_1}, \qquad U_{II}^2 = \frac{Q_c}{\kappa_2}, \qquad K^2 = \frac{Q_r^2}{\kappa_1 \kappa_2}.
$$

Thus, the wavespeeds are given by

$$
U_N^2 = \frac{Q_{ij} n_i n_j}{\rho_0}, \tag{9.14}
$$

and from (9.13),

$$
U_N^2 = \frac{1}{2}\{U_I^2 + U_{II}^2 \pm \sqrt{(U_I^2 + U_{II}^2)^2 - 4(U_I^2 U_{II}^2 - K^2)}\}. \tag{9.15}
$$

To understand the last equation we note

$$
\begin{aligned}
U_I^2 U_{II}^2 - K^2 = \frac{\rho_0}{\kappa_1 \kappa_2}\Big\{ & N_A N_B \frac{\partial^2 \psi}{\partial \nu_{,A} \partial \nu_{,B}} N_C N_D \frac{\partial^2 \psi}{\partial \omega_{,C} \partial \omega_{,D}} \\
& - \Big(N_A N_B \frac{\partial^2 \psi}{\partial \omega_{,A} \partial \nu_{,B}} \Big)^2 \Big\}.
\end{aligned}
\tag{9.16}
$$

We believe that the condition that the right hand side of (9.16) is positive is a physical one which for example, in the linear theory is a condition required for uniqueness in the dynamic theory when using an energy method. Hence, we believe $U_I^2 U_{II}^2 - K^2 > 0$ is a reasonable assumption and so for the centrosymmetric case we have an elastic wave with wavespeed given by (9.14) and a fast and a slow wave with wavespeeds given by (9.15).

9.1.1 Amplitude Equation

We develop the amplitude equation in the one-dimensional case. This result is essentially the same as one finds for a *plane* acceleration wave moving in a three-dimensional body.

In terms of the displacement u, in one dimension the governing equations for a double distribution of voids in a nonlinear isothermal elastic body may be deduced from (9.1) - (9.3) and they are

$$\begin{aligned} \rho_0 \ddot{u} &= \Pi_X\,, \\ \kappa_1 \ddot{\nu} &= h_X + g, \\ \kappa_2 \ddot{\omega} &= j_X + \hat{h}, \end{aligned} \tag{9.17}$$

where h and j are the fluxes and g and $\hat{h}$ now correspond to the terms g and h in (9.2) and (9.3). Since the functions Π, h, g, j and $\hat{h}$ depend on $F = u_X, \nu, \omega, \nu_X$ and ω_X we may expand (9.17) to find

$$\begin{aligned} \rho_0 \ddot{u} &= \Pi_F u_{XX} + \Pi_\nu \nu_X + \Pi_\omega \omega_X + \Pi_{\nu_X} \nu_{XX} + \Pi_{\omega_X} \omega_{XX}\,, \\ \kappa_1 \ddot{\nu} &= h_F u_{XX} + h_\nu \nu_X + h_\omega \omega_X + h_{\nu_X} \nu_{XX} + h_{\omega_X} \omega_{XX} + g, \\ \kappa_2 \ddot{\omega} &= j_F u_{XX} + j_\nu \nu_X + j_\omega \omega_X + j_{\nu_X} \nu_{XX} + j_{\omega_X} \omega_{XX} + \hat{h}. \end{aligned} \tag{9.18}$$

Define the amplitudes a, b and c by

$$a(t) = [\ddot{u}], \qquad b(t) = [\ddot{\nu}], \qquad c(t) = [\ddot{\omega}],$$

and then taking the jumps of (9.18) and using the Helmholtz relation to see, for example,

$$[\ddot{u}] = a = V^2 [u_{XX}],$$

with V being the wavespeed one may show

$$\begin{aligned} (\Pi_F - \rho_0 V^2) a + \Pi_{\nu_X} b + \Pi_{\omega_X} c &= 0, \\ h_F a + (h_{\nu_X} - \kappa_1 V^2) b + h_{\omega_X} c &= 0, \\ j_F a + j_{\nu_X} b + (j_{\omega_X} - \kappa_2 V^2) c &= 0. \end{aligned} \tag{9.19}$$

Since we require $a, b, c \neq 0$, system (9.19) leads to the wavespeed equation

$$
\begin{aligned}
&(\rho_0 V^2 - \Pi_F)(\kappa_1 V^2 - h_{\nu_X})(\kappa_2 V^2 - j_{\omega_X}) - \Pi_{\omega_X} j_F(\kappa_1 V^2 - h_{\nu_X}) \\
&\quad - \Pi_{\nu_X} h_F(\kappa_2 V^2 - j_{\omega_X}) - h_{\omega_X} j_{\nu_X}(\rho_0 V^2 - \Pi_F) \\
&\quad - \Pi_{\nu_X} h_{\omega_X} j_F - \Pi_{\omega_X} h_F j_{\nu_X} = 0.
\end{aligned} \tag{9.20}
$$

To find the amplitude equation we suppose the wave is moving into an equilibrium region where

$$
u_X = \text{constant}, \qquad \nu = \text{constant}, \qquad \omega = \text{constant}. \tag{9.21}
$$

This state requires the functions g and $\hat{h}$ to be such that

$$
g|_E = 0, \qquad \hat{h}|_E = 0,
$$

where E denotes equilibrium. These conditions are not vacuous and lead to equilibrium conditions between u_x, ω and ν once the form for ψ is known. For example, in the linear case we find from (3.69) that in equilibrium

$$
\begin{aligned}
\alpha_1 \nu + \alpha_3 \omega + b_{ij} u_{i,j} &= 0, \\
\alpha_2 \omega + \alpha_3 \nu + d_{ij} u_{i,j} &= 0.
\end{aligned}
$$

The technique to derive the amplitude equation is to differentiate each of (9.18) with respect to X and then take the jumps of the results, recalling (9.21). This leads to the equations

$$
\begin{aligned}
\rho_0[\ddot{u}_X] =& \Pi_F[u_{XXX}] + \Pi_{FF}[u_{XX}^2] + \Pi_\nu[\nu_{XX}] + \Pi_\omega[\omega_{XX}] \\
&+ 2\Pi_{F\nu_X}[u_{XX}\nu_{XX}] + 2\Pi_{F\omega_X}[u_{XX}\omega_{XX}] + 2\Pi_{\nu_X\omega_X}[\nu_{XX}\omega_{XX}] \\
&+ \Pi_{\nu_X\nu_X}[\nu_{XX}^2] + \Pi_{\omega_X\omega_X}[\omega_{XX}^2] + \Pi_{\nu_X}[\nu_{XXX}] + \Pi_{\omega_X}[\omega_{XXX}],
\end{aligned} \tag{9.22}
$$

and

$$
\begin{aligned}
\kappa_1[\ddot{\nu}_X] =& h_F[u_{XXX}] + h_{FF}[u_{XX}^2] + h_\nu[\nu_{XX}] + h_\omega[\omega_{XX}] \\
&+ 2h_{F\nu_X}[u_{XX}\nu_{XX}] + 2h_{F\omega_X}[u_{XX}\omega_{XX}] + 2h_{\nu_X\omega_X}[\nu_{XX}\omega_{XX}] \\
&+ h_{\nu_X\nu_X}[\nu_{XX}^2] + h_{\omega_X\omega_X}[\omega_{XX}^2] + h_{\nu_X}[\nu_{XXX}] + h_{\omega_X}[\omega_{XXX}] \\
&+ g_F[u_{XX}] + g_{\nu_X}[\nu_{XX}] + g_{\omega_X}[\omega_{XX}],
\end{aligned} \tag{9.23}
$$

and

$$
\begin{aligned}
\kappa_2[\ddot{\omega}_X] =& j_F[u_{XXX}] + j_{FF}[u_{XX}^2] + j_\nu[\nu_{XX}] + j_\omega[\omega_{XX}] \\
&+ 2j_{F\nu_X}[u_{XX}\nu_{XX}] + 2j_{F\omega_X}[u_{XX}\omega_{XX}] + 2j_{\nu_X\omega_X}[\nu_{XX}\omega_{XX}] \\
&+ j_{\nu_X\nu_X}[\nu_{XX}^2] + j_{\omega_X\omega_X}[\omega_{XX}^2] + j_{\nu_X}[\nu_{XXX}] + j_{\omega_X}[\omega_{XXX}] \\
&+ \hat{h}_F[u_{XX}] + \hat{h}_{\nu_X}[\nu_{XX}] + \hat{h}_{\omega_X}[\omega_{XX}].
\end{aligned} \tag{9.24}
$$

We now use the Hadamard relation to see that

$$
\begin{aligned}
-[\ddot{u}_X] &= \frac{2}{V}\frac{\delta a}{\delta t} - V^2[u_{XXX}],\\
-[\ddot{v}_X] &= \frac{2}{V}\frac{\delta b}{\delta t} - V^2[v_{XXX}],\\
-[\ddot{\omega}_X] &= \frac{2}{V}\frac{\delta c}{\delta t} - V^2[\omega_{XXX}].
\end{aligned}
\tag{9.25}
$$

Expressions (9.25) are employed in equations (9.22) - (9.24) together with the jump of a product (8.44). We form what results from equations (9.22), (9.23) and (9.24) into the combination

$$(9.22)+\lambda_1(9.23)+\lambda_2(9.24).$$

The coefficients λ_1 and λ_2 are chosen to ensure the terms involving $[u_{XXX}],[v_{XXX}]$ and $[\omega_{XXX}]$ disappear. This also requires use of the wavespeed conditions (9.20). The appropriate choice of λ_1 and λ_2 is

$$
\begin{aligned}
\lambda_1 &= \frac{1}{D}\{j_{v_X}(\rho_0 V^2 - \Pi_F) + j_F \Pi_{v_X}\},\\
\lambda_2 &= \frac{1}{D}\{(\kappa_1 V^2 - h_{v_X})(\rho_0 V^2 - \Pi_F) - h_F \Pi_{v_X}\},
\end{aligned}
\tag{9.26}
$$

where the denominator D is given by the expression

$$D = \frac{\partial h}{\partial F}\frac{\partial j}{\partial v_X} + \frac{\partial j}{\partial F}\left(\kappa_1 V^2 - \frac{\partial h}{\partial v_X}\right).$$

The resulting single equaton involves a, b and c. We eliminate two of these variables using (9.19). For example if we eliminate b and c we can write these variables in terms of a by using the relation involving a matrix acting on a vector, namely

$$
\begin{pmatrix} b \\ c \end{pmatrix} = \frac{1}{\Delta}
\begin{pmatrix} \kappa_2 V^2 - \dfrac{\partial j}{\partial \omega_X} & \dfrac{\partial h}{\partial \omega_X} \\[2ex] \dfrac{\partial j}{\partial v_X} & \kappa_1 V^2 - \dfrac{\partial h}{\partial v_X} \end{pmatrix}
\begin{pmatrix} \dfrac{\partial h}{\partial F} a \\[2ex] \dfrac{\partial j}{\partial F} a \end{pmatrix}
\tag{9.27}
$$

where

$$\Delta = \left(\kappa_1 V^2 - \frac{\partial h}{\partial v_X}\right)\left(\kappa_2 V^2 - \frac{\partial j}{\partial \omega_X}\right) - \frac{\partial h}{\partial \omega_X}\frac{\partial j}{\partial v_X}.$$

In this way one shows $a(t)$ satsifies an equation of form

$$\frac{\delta a}{\delta t} + \mu_1 a^2 + \mu_2 a = 0, \tag{9.28}$$

where since the wave is moving into equilibrium μ_1 and μ_2 are constants. In fact, this equation has precise form

$$\begin{aligned}
&2\rho_0 V\frac{\delta a}{\delta t}+2\lambda_1\kappa_1 V\frac{\delta b}{\delta t}+2\lambda_2\kappa_2 V\frac{\delta c}{\delta t}\\
&+\frac{a^2}{V^2}(\Pi_{FF}+\lambda_1 h_{FF}+\lambda_2 j_{FF})+2\frac{ab}{V^2}(\Pi_{F\nu_X}+\lambda_1 h_{F\nu_X}+\lambda_2 j_{F\nu_X})\\
&+2\frac{ac}{V^2}(\Pi_{F\omega_X}+\lambda_1 h_{F\omega_X}+\lambda_2 j_{F\omega_X})+\frac{b^2}{V^2}(\Pi_{\nu_X\nu_X}+\lambda_1 h_{\nu_X\nu_X}+\lambda_2 j_{\nu_X\nu_X})\\
&+2\frac{bc}{V^2}(\Pi_{\nu_X\omega_X}+\lambda_1 h_{\nu_X\omega_X}+\lambda_2 j_{\nu_X\omega_X})\\
&+\frac{c^2}{V^2}(\Pi_{\omega_X\omega_X}+\lambda_1 h_{\omega_X\omega_X}+\lambda_2 j_{\omega_X\omega_X})\\
&+a(\lambda_1 g_F+\lambda_2\hat{h}_F)+b(h_\nu+g_{\nu_X}+j_\nu+\hat{h}_{\nu_X})\\
&+c(\Pi_\omega+h_\omega+g_{\omega_X}+j_\omega+\hat{h}_{\omega_X})=0.
\end{aligned} \tag{9.29}$$

Once (9.27) is used to substitute for b abd c, and (9.26) is employed to substitute for λ_1 and λ_2, (9.29) leads to (9.28). This is a situation where a computer algebra package will help to yield the coefficients μ_1 and μ_2 and it shows that even with only two voids the calculation of coefficients becomes very involved.

Equation (9.28) is a Bernoulli equation with constant coefficients and is easily solved as in (8.46). The solution yields the complete amplitude behaviour and finite time blow-up is to be expected under the right initial conditions.

9.2 Thermoelastic Acceleration Waves

In this section we study an acceleration wave for a thermoelastic body with a double porosity (void distribution). The notation and equations begin as in section 9.1, so we have equations (9.1), (9.2) and (9.3) together with the relations (9.4). In addition, however, we must add the balance of energy equation (3.66) so that with the heat source set to zero

$$\rho_0\theta\dot{\eta}=-q_{A,A}. \tag{9.30}$$

The Helmholtz free energy and the heat flux now satisfy the constitutive theory

$$\psi=\psi(F_{iA},\nu,\omega,\nu_{,A},\omega_{,A},\theta), \tag{9.31}$$

and

$$q_A=q_A(F_{iA},\nu,\omega,\nu_{,A},\omega_{,A},\theta,\theta_{,A}). \tag{9.32}$$

An acceleration wave in a nonlinear thermoelastic body with double porosity is defined to be a singular surface $\mathscr{S}$ such that u_i,ν,ω,θ are $C^1(\mathbb{R}^3\times\{t>0\})$ but $\ddot{u}_i,\dot{u}_{i,A},u_{i,AB}$, $\ddot{\nu},\dot{\nu}_{,A},\nu_{,AB}$, $\ddot{\omega},\dot{\omega}_{,A},\omega_{,AB}$, $\ddot{\theta},\dot{\theta}_{,A}$ and $\theta_{,AB}$ and their higher derivatives suffer a finite discontinuity across $\mathscr{S}$. To find the wavespeeds we use the chain rule on equations (9.1), (9.2), (9.3) and (9.30) employing (9.4) together with (9.31) and (9.32). We then take the jumps of the four resulting equations to find that

$$
\begin{aligned}
\rho_0[\ddot{u}_i] &= \frac{\partial \Pi_{Ai}}{\partial F_{jB}}[u_{j,BA}] + \frac{\partial \Pi_{Ai}}{\partial \nu_{,K}}[\nu_{,KA}] + \frac{\partial \Pi_{Ai}}{\partial \omega_{,K}}[\omega_{,KA}],\\
\kappa_1[\ddot{\nu}] &= \frac{\partial h_A}{\partial F_{jB}}[u_{j,BA}] + \frac{\partial h_A}{\partial \nu_{,K}}[\nu_{,KA}] + \frac{\partial h_A}{\partial \omega_{,K}}[\omega_{,KA}],\\
\kappa_2[\ddot{\omega}] &= \frac{\partial j_A}{\partial F_{jB}}[u_{j,BA}] + \frac{\partial j_A}{\partial \nu_{,K}}[\nu_{,KA}] + \frac{\partial j_A}{\partial \omega_{,K}}[\omega_{,KA}],\\
\rho_0\theta&\left\{\frac{\partial \eta}{\partial F_{iA}}[\dot{u}_{i,A}] + \frac{\partial \eta}{\partial \nu_{,A}}[\dot{\nu}_{,A}] + \frac{\partial \eta}{\partial \omega_{,A}}[\dot{\omega}_{,A}]\right\}\\
&= -\frac{\partial q_A}{\partial F_{iB}}[u_{i,BA}] - \frac{\partial q_A}{\partial \theta_{,B}}[\theta_{,AB}] - \frac{\partial q_A}{\partial \nu_{,B}}[\nu_{,AB}] - \frac{\partial q_A}{\partial \omega_{,B}}[\omega_{,AB}].
\end{aligned}
\tag{9.33}
$$

Define the wave amplitudes a_i, b, c and f by

$$
a_i(t) = [\ddot{u}_i], \qquad b(t) = [\ddot{\nu}], \qquad c(t) = [\ddot{\omega}], \qquad f(t) = [\ddot{\theta}].
$$

Note that f appears only in $(9.33)_4$. Therefore, the amplitudes a_i, b and c still satisfy the equations (9.7). The comments following equation (9.10) regarding longitudinal and transverse waves and centrosymmetric bodies still apply here, as do the comments regarding wavesppeds. The difference here is that the coefficients now depend on θ and $\theta_{,A}$.

We may employ the Hadamard relation and the compatibility conditions to see that

$$
\begin{aligned}
&[\dot{u}_{i,A}] = -U_N[N^B u_{i,AB}], \qquad [\ddot{u}_i] = U_N^2[N^A N^B u_{i,AB}],\\
&[u_{i,AB}] = N_A N_B[u_{i,KL}N^K N^L] = \frac{N_A N_B}{U_N^2}\,a_i,\\
&[\dot{u}_{i,A}] = -U_N N_A[N^B N^C u_{i,CB}] = -\frac{1}{U_N}N_A a_i,
\end{aligned}
\tag{9.34}
$$

together with similar relations for ν, ω and θ. Applying these relations to equation $(9.33)_4$ one may derive

$$
\begin{aligned}
\frac{\partial q_A}{\partial \theta_{,B}} N_A N_B f =& \left(\rho_0\theta N_A U_N \frac{\partial \eta}{\partial F_{iA}} - \frac{\partial q_A}{\partial F_{iB}} N_A N_B\right) a_i\\
&+ \left(\rho_0\theta N_A U_N \frac{\partial \eta}{\partial \nu_{,A}} - \frac{\partial q_A}{\partial \nu_{,B}} N_A N_B\right) b\\
&+ \left(\rho_0\theta N_A U_N \frac{\partial \eta}{\partial \omega_{,A}} - \frac{\partial q_A}{\partial \omega_{,B}} N_A N_B\right) c.
\end{aligned}
\tag{9.35}
$$

Once a_i, b and c are known, equation (9.35) yields the temperature amplitude $f(t)$.

One may progress and calculate the amplitudes and then determine explicitly a_i, b, c and f, see e.g. exercise 9.3.

9.3 Comparison of Nonlinear Waves

In chapter 4 we compared theories of elastic bodies with pores to those of elastic bodies with voids by analysing uniqueness and linear wave motion. We make a similar comparison here for nonlinear acceleration waves in the double porosity and double voids theories.

With the extended Biot theory developed in section 8.2.1 we see that there is a pair of acceleration waves, one which moves to the right and the other to the left, with the same wavespeed U_N determined by equation (8.30). This equation involves directly the elastic wave amplitude a_i. This amplitude may be calculated as in section 8.2.2. Once the amplitude a_i is known the amplitudes associated with the macro pressure and micro pressure, namely b and c, follow directly from equations (8.33) and (8.34).

Acceleration waves in the nonlinear double voids theory are studied in section 9.1 and these do not lead to a single wavespeed equation. Instead we find a coupled system of equations for the elastic wave amplitude a_i, the macro void wave amplitude b, and the micro void wave amplitude c, namely equations (9.7). These equations lead to a wavespeed equation, (9.11), which allows for three pairs of coupled waves, each pair consisting of a left and a right moving wave with the same speed, although the speed of each of the pairs is different. This is a more complicated mathematical structure than in the double porosity case since it leads to an elastic wave, a wave associated to the macro void distribution, and a wave associated to the micro void distribution. In general, the waves are interconnected and the wavespeeds are determined from the sixth order equation (9.11). The wave amplitudes may be calculated as in section 9.1.1 but the calculation is more involved than in the double porosity case, although the solution of each amplitude still reduces to solving a Bernoulli equation.

9.4 Exercises

Exercise 9.1. For the nonlinear single void theory equations (3.38), (3.39), develop a three-dimensional acceleration wave analysis and calculate the equations governing the wavespeeds, cf. Iesan [111], Straughan [202], pp. 312, 313.

Exercise 9.2. For the one-dimensional nonlinear single void theory equations (3.38), (3.39), in one space dimension, calculate the amplitude equation for an acceleration wave.

Exercise 9.3. For the one-dimensional version of the double voids thermoelasticity theory given by equations (9.1), (9.2), (9.3) and (9.30) calculate the amplitude equation for an acceleration wave.

Chapter 10
Double Porosity and Second Sound

Up to this point in this book we have considered thermoelastic bodies with a multiple porosity structure assuming that the heat flux is governed by a Fourier law, i.e. the heat flux is determined directly by the temperature gradient. Materials technology is advancing very rapidly and it is likely that for many future applications a Fourier law may be inadequate. The phenomenon of temperature travelling as a wave is studied in detail in the book by Straughan [203] and he outlines many examples of classical elasticity where the temperature field has wave like behaviour. This behaviour is incorporated by a variety of mechanisms such as Cattaneo theory, Green-Laws $\theta, \dot{\theta}$ theory, Green-Naghdi type II or type III theory, heat flux delay, and others, see Straughan [203] for details, see also Jou *et al.* [130]. This area of heat transfer is still very active in modern continuum mechanics and engineering, see e.g. Alharbi & Scott [2, 3], Avalishvili *et al.* [7], Aouadi *et al.* [5], Barbosa *et al.* [11], Bissell [20, 21], Both *et al.* [26], Chirita *et al.* [41], Christov [46], Christov & Jordan [48], Coti Zelati *et al.* [56], Dell'Oro & Pata [68], De Sciarra & Salerno [64], Eltayeb [72], Fabrizio [74], Fabrizio *et al.* [76], Fabrizio & Franchi [75], Fabrizio & Lazzari [79], Fatori *et al.* [80], Fernandez & Masid [82], Franchi *et al.* [86], Giorgi *et al.* [93], Hayat *et al.* [101], Hayat *et al.* [102], Jordan [123], Jordan *et al.* [127], Jou *et al.* [130], Jou *et al.* [128, 129], Jou & Cimmelli [131], Khayat *et al.* [135], Kuang [147], Leseduarte & Quintanilla [151], Liu [155], Liu *et al.* [153, 154], Mustafa [164], Quintanilla & Sivaloganathan [176], Randrianalisoa *et al.* [177], Rukolaine [182], Sabelnikov *et al.* [183], Sellitto *et al.* [191], Singh & Tadmor [197], Straughan [203], Sui *et al.* [210], Van [222], Wakeni *et al.* [224]. It is natural, therefore, to consider theories of thermoelasticity with multiple porosity but where the thermodynamics allows one to incorporate heat transfer by second sound, i.e. where heat may travel directly as a wave. In addition to newly fabricated materials such multi-porosity thermoelasticity with second sound may well be important in materials with phase change effects, or where damage is present. These theories are difficult even in classical thermodynamics, see e.g. Amendola & Fabrizio [4], Berti *et al.* [17], Berti *et al.* [16], Bonetti *et al.* [22], Caputo & Fabrizio [28], Cavaterra *et al.* [31], Fabrizio *et al.* [78, 77]. However, their description will present an interesting challenge.

B. Straughan, *Mathematical Aspects of Multi–Porosity Continua*, Advances in Mechanics and Mathematics 38, https://doi.org/10.1007/978-3-319-70172-1_10

In this chapter we begin a development of thermoelasticity with a double porosity structure where we allow for temperature to travel as a wave. We do this by developing two new theories. The first is to generalize the Svanadze equations (2.34) and incorporate second sound by a generalized Cattaneo theory, cf. Caviglia *et al.* [32], Straughan [203], equations (2.39). The second development will extend the Iesan & Quintanilla [115] theory by incorporating a Green & Naghdi [97] type II thermodynamics.

10.1 Cattaneo Theory and Double Porosity Elasticity

We commence with the Svanadze equations (2.34) for a thermoelastic body with a double porosity structure governed by the macro pressure, p, and the micro pressure, q. Equations $(2.34)_{1,2,3}$ remain the same but equation $(2.34)_4$ is replaced by

$$a\dot{\theta} + a_{ij}\dot{u}_{i,j} = -q_{i,i} + \rho r, \tag{10.1}$$

where q_i is the heat flux. In equation $(2.34)_4$ the heat flux is governed by the Fourier law $q_i = -\kappa_{ij}\theta_{,j}$. Here we adopt instead a Cattaneo approach and write

$$\tau\dot{q}_i + q_i = -\kappa_{ij}\theta_{,j}\,. \tag{10.2}$$

The reasons for adopting equation (10.2) are explained in depth in chapter 1 of the book by Straughan [203]. The tensor κ_{ij} is positive-definite and we denote its inverse tensor by c_{ij}. We rewrite equation (10.2) in terms of c_{ij} as

$$\tau c_{ij}\dot{q}_j = -c_{ij}q_j - \theta_{,i}\,. \tag{10.3}$$

Hence, the full system of equations governing the behaviour of the thermoelastic model with double porosity is

$$\begin{aligned}
\rho\ddot{u}_i &= (a_{ijkh}u_{k,h})_{,j} - (\beta_{ij}p)_{,j} - (\gamma_{ij}q)_{,j} - (a_{ij}\theta)_{,j} + \rho f_i\,,\\
\alpha\dot{p} &= (k_{ij}p_{,j})_{,i} - \gamma(p-q) - \beta_{ij}\dot{u}_{i,j} + \rho s_1\,,\\
\beta\dot{q} &= (m_{ij}q_{,j})_{,i} + \gamma(p-q) - \gamma_{ij}\dot{u}_{i,j} + \rho s_2\,,\\
a\dot{\theta} &= -a_{ij}\dot{u}_{i,j} - q_{i,i} + \rho r\,,\\
\tau c_{ij}\dot{q}_j &= -c_{ij}q_j - \theta_{,i}\,.
\end{aligned} \tag{10.4}$$

It is important to note that Caviglia *et al.* [32] show that in an elastic body a_{ijkh} can incorporate a pre-stress. As Straughan [203], p. 47, points out, this means that it is unlikely that the elastic coefficients a_{ijkh} will in general be sign-definite.

In the linear case like (10.4) one may also think of this as being a Green-Naghdi type III model for a thermoelastic body with double porosity. To see this in a heuristic manner we consider equations $(10.4)_{4,5}$ with $r = 0$ and with the term $a_{ij}\dot{u}_{i,j}$ missing. Then we have

$$
\begin{aligned}
a\dot{\theta} &= -q_{i,i}\,, \\
\tau\dot{q}_i &= -q_i - \kappa_{ij}\theta_{,j}\,,
\end{aligned}
\tag{10.5}
$$

where we have employed the inverse to c_{ij}. Upon elimination of q_i these equations lead to

$$
a\tau\ddot{\theta} + a\dot{\theta} = (k_{ij}\theta_{,j})_{,i} \tag{10.6}
$$

which is a damped wave equation, cf. the type III equation in Green & Naghdi [97]. Jordan [120] essentially argues that one can achieve a type II heat transfer theory by replacing $(10.5)_2$ by an equation of form

$$
\dot{q}_i = -\hat{\kappa}_{ij}\theta_{,j}\,. \tag{10.7}
$$

Coupled with $(10.5)_1$ this then leads to an equation of the type

$$
a_1\ddot{\theta} = (\hat{k}_{ij}\theta_{,j})_{,i} \tag{10.8}
$$

for a suitable coefficient a_1. This is a dissipationless form of heat wave equation, cf. Green & Naghdi [97] type II theory, and see the development from the entropy balance equation for type II theory for a rigid body in Straughan [203], pp. 29–31.

In view of the above arguments one may then argue for an alternative thermoelastic double porosity theory based essentially on (10.7) and $(10.4)_4$. In this case instead of system (10.4) we here omit the q_i term in $(10.5)_2$ and propose the partial differential equations

$$
\begin{aligned}
\rho\ddot{u}_i &= (a_{ijkh}u_{k,h})_{,j} - (\beta_{ij}p)_{,j} - (\gamma_{ij}q)_{,j} - (a_{ij}\theta)_{,j} + \rho f_i\,, \\
\alpha\dot{p} &= (k_{ij}p_{,j})_{,i} - \gamma(p-q) - \beta_{ij}\dot{u}_{i,j} + \rho s_1\,, \\
\beta\dot{q} &= (m_{ij}q_{,j})_{,i} + \gamma(p-q) - \gamma_{ij}\dot{u}_{i,j} + \rho s_2\,, \\
a\dot{\theta} &= -a_{ij}\dot{u}_{i,j} - q_{i,i} + \rho r\,, \\
\tau c_{ij}\dot{q}_j &= -\theta_{,i}\,.
\end{aligned}
\tag{10.9}
$$

10.1.1 Uniqueness and Continuous Dependence

We now establish uniqueness and continuous dependence for a solution to equations (10.4) on a bounded domain $\Omega \subset \mathbb{R}^3$, $t > 0$, with boundary Γ. We suppose that $\beta_{ij}, \gamma_{ij}, a_{ij}$, k_{ij}, m_{ij}, c_{ij} are symmetric tensors with k_{ij}, m_{ij}, a_{ij} and c_{ij} positive - definite. In addition a_{ijkh} satisfy the symmetries

$$
a_{ijkh} = a_{jikh} = a_{khij}
$$

and are positive - definite so that

$$
a_{ijkh}\xi_{ij}\xi_{kh} \geq a_0\xi_{ij}\xi_{ij}\,, \qquad \forall\, \xi_{ij}\,.
$$

for some constant $a_0 > 0$. The coefficients ρ,α,β,a and γ may all depend on $\mathbf{x}$ but are always positive. The boundary condiitons are

$$\begin{aligned} &u_i(\mathbf{x},t) = u_i^B(\mathbf{x},t), \quad p(\mathbf{x},t) = p^B(\mathbf{x},t), \\ &q(\mathbf{x},t) = q^B(\mathbf{x},t), \quad \theta(\mathbf{x},t) = \theta^B(\mathbf{x},t), \qquad \text{on } \Gamma \times \{t > 0\}, \end{aligned} \tag{10.10}$$

and the initial conditions are

$$\begin{aligned} &u_i(\mathbf{x},0) = v_i(\mathbf{x}), \quad \dot{u}_i(\mathbf{x},0) = w_i(\mathbf{x}), \\ &p(\mathbf{x},0) = P(\mathbf{x}), \quad q(\mathbf{x},0) = Q(\mathbf{x}), \\ &\theta(\mathbf{x},0) = R(\mathbf{x}), \quad q_i(\mathbf{x},0) = Q_i(\mathbf{x}), \qquad \mathbf{x} \in \Omega, \end{aligned} \tag{10.11}$$

for prescribed functions $u_i^B,\dots,Q_i$. Denote the boundary-initial value problem given by (10.4), (10.10) and (10.11) by $\mathscr{P}$.

Let $(u_i^1,p^1,q^1,\theta^1,q_i^1)$ and $(u_i^2,p^2,q^2,\theta^2,q_i^2)$ be two solutions to $\mathscr{P}$ for the same boundary conditions, the same initial conditions, and the same source terms f_i,s_1,s_2 and r. Then the difference solution $u_i = u_i^1 - u_i^2$, $p = p^1 - p^2$, $q = q^1 - q^2$, $\theta = \theta^1 - \theta^2$ and $q_i = q_i^1 - q_i^2$ satisfies equations (10.4) with f_i,s_1,s_2 and r removed, together with zero boundary conditions and zero initial conditions.

To establish uniqueness multiply $(10.4)_{1-5}$ in turn by $\dot{u}_i$, p,q,θ and q_i, respectively. Integrate each resulting equation over Ω to find after some integration by parts and use of the boundary conditions,

$$\begin{aligned} &\frac{d}{dt}\left[\frac{1}{2} < \rho\dot{u}_i\dot{u}_i > + \frac{1}{2} < a_{ijkh}u_{i,j}u_{k,h} >\right] \\ &\qquad = < \beta_{ij}p\dot{u}_{i,j} > + < \gamma_{ij}q\dot{u}_{i,j} > + < a_{ij}\theta\dot{u}_{i,j} >, \\ &\frac{d}{dt}\frac{1}{2} < \alpha p^2 > + < k_{ij}p_{,i}p_{,j} > + < \gamma p(p-q) > = - < \beta_{ij}\dot{u}_{i,j}p >, \\ &\frac{d}{dt}\frac{1}{2} < \beta q^2 > + < m_{ij}q_{,i}q_{,j} > - < \gamma p(p-q) > = - < \gamma_{ij}\dot{u}_{i,j}q >, \\ &\frac{d}{dt}\frac{1}{2} < a\theta^2 > = - < a_{ij}\dot{u}_{i,j}\theta > + < q_i\theta_{,i} >, \\ &\frac{d}{dt}\frac{1}{2} < \tau c_{ij}q_iq_j > + < c_{ij}q_iq_j > = - < q_i\theta_{,i} > . \end{aligned}$$

Add the five above equations and then integrate over $(0,t)$ to obtain

$$\begin{aligned} E(t) &+ \int_0^t < k_{ij}p_{,i}p_{,j} > ds + \int_0^t < m_{ij}q_{,i}q_{,j} > ds \\ &+ \int_0^t < \gamma(p-q)^2 > ds + \int_0^t < c_{ij}q_iq_j > ds = E(0), \end{aligned} \tag{10.12}$$

where

$$
\begin{aligned}
E(t) =& \frac{1}{2} < \rho \dot{u}_i \dot{u}_i > + \frac{1}{2} < a_{ijkh} u_{i,j} u_{k,h} > + \frac{1}{2} < \alpha p^2 > \\
&+ \frac{1}{2} < \beta q^2 > + \frac{1}{2} < a\theta^2 > + \frac{1}{2} < \tau c_{ij} q_i q_j > .
\end{aligned}
$$

For uniqueness $E(0) = 0$ and then using the positivity conditions on the coefficients we see

$$
0 \le E(t) \le 0
$$

and so uniqueness follows.

One may establish continuous dependence upon the initial data from (10.12) when the initial data are taken to be non-zero. Equation (10.12) leads to continuous dependence in the $L^2(\Omega)$ norm for p, q, θ and q_i, and in the $H_0^1(\Omega)$ norm for u_i.

10.1.2 Acceleration - Temperature Waves

We now show how one may study the motion of an acceleration wave for the Jordan version of the Cattaneo double porosity elasticity model given by equations (10.9).

We consider equations (10.9) with zero source terms, i.e. $f_i = 0, s_1 = s_2 = 0$ and $r = 0$. Further, to avoid confusion with using a_i as the amplitude we replace the coefficient of $\dot{\theta}$ by δ rather than a. Thus, the system of equations under consideration is

$$
\begin{aligned}
\rho \ddot{u}_i &= (a_{ijkh} u_{k,h})_{,j} - (\beta_{ij} p)_{,j} - (\gamma_{ij} q)_{,j} - (a_{ij}\theta)_{,j} + \rho f_i , \\
\alpha \dot{p} &= (k_{ij} p_{,j})_{,i} - \gamma(p - q) - \beta_{ij} \dot{u}_{i,j} + \rho s_1 , \\
\beta \dot{q} &= (m_{ij} q_{,j})_{,i} + \gamma(p - q) - \gamma_{ij} \dot{u}_{i,j} + \rho s_2 , \\
\delta \dot{\theta} &= -a_{ij} \dot{u}_{i,j} - q_{i,i} + \rho r , \\
\tau c_{ij} \dot{q}_j &= -\theta_{,i} .
\end{aligned}
\tag{10.13}
$$

We define an acceleration wave for (10.13) to be a surface $\mathscr{S}$ such that $u_i \in C^1(\mathbb{R}^3 \times \{t > 0\}), p, q, \theta, q_i \in C(\mathbb{R}^3 \times \{t > 0\})$, but across $\mathscr{S}$, $\ddot{u}_i, \dot{u}_{i,j}, u_{i,jk}\ \dot{p}, p_{,i}, \dot{q}, q_{,i}$, $\dot{\theta}, \theta_{,i}$, $\dot{q}_i$ and $q_{i,j}$ and their higher derivatives suffer a finite discontinuity.

Recalling the definition of an acceleration wave we take the jumps of equations (10.13) to see that

$$
\begin{aligned}
\rho [\ddot{u}_i] &= a_{ijkh} [u_{k,hj}] - a_{ij} [\theta_{,j}] , \\
\delta [\dot{\theta}] &= -a_{ij} [\dot{u}_{i,j}] - [q_{i,i}] , \\
\tau [\dot{q}_i] &= -\kappa_{ij} [\theta_{,j}] ,
\end{aligned}
\tag{10.14}
$$

and

$$
\begin{aligned}
k_{ij} [p_{,ij}] &= \beta_{ij} [\dot{u}_{i,j}] , \\
m_{ij} [q_{,ij}] &= \gamma_{ij} [\dot{u}_{i,j}] .
\end{aligned}
\tag{10.15}
$$

In deriving $(10.14)_3$ we have used $\kappa_{ij} = c_{ij}^{-1}$ in (10.9).

The wave amplitudes $a_i, \mathscr{T}, Q_i, P$ and Q are defined by

$$a_i(t)=[\ddot{u}_i], \qquad \mathscr{T}(t)=[\dot{\theta}], \qquad Q_i(t)=[\dot{q}_i], \qquad P(t)=[\dot{p}], \qquad Q(t)=[\dot{q}].$$

The reason for separating equations (10.14) and (10.15) is that (10.14) yield a coupled system involving $a_i, \mathscr{T}$ and Q_i whereas (10.15) show that the void wave amplitudes P and Q play no role in the wavespeed analysis and P and Q follow once $a_i, \mathscr{T}$ and Q_i are determined. We employ the compatibility and Hadamard relations (4.11) and (4.10) to derive relations like (4.12), (4.13) and (4.14) together with the relations

$$[\dot{\theta}]=-U_N[n_a\theta_{,a}], \qquad [\theta_{,j}]=n_j[\theta_{,a}n_a]$$

and from (10.14) we may show that

$$(\rho U_N^2\delta_{ij}-Q_{ij})a_j=a_{ij}n_jU_N\mathscr{T}, \tag{10.16}$$

in addition to

$$-a_{ij}n_ja_i+\delta U_N\mathscr{T}-Q_in_i=0, \tag{10.17}$$

together with

$$\tau U_NQ_i=\kappa_{ij}n_j\mathscr{T}, \tag{10.18}$$

where Q_{ij} is the acoustic tensor $Q_{ij}=a_{irjs}n_rn_s$.

From equation (10.16) we may use theorem 2 of Chadwick & Currie [33] to infer there is at least one direction n_i^* such that $a_{ij}n_j^*$ is an eigenvector of the acoustic tensor Q_{ij}. If the region ahead of the wave is at rest in a homogeneous equilibrium configuration then a_{ij} is constant and so n_i^* is fixed. Thus, we may infer the existence of a plane acceleration wave moving in the direction of n_i^* which has its amplitude in the direction $a_{ij}n_j^*$. Such a wave may be referred to as a generalized longitudinal wave. Let λ_i be the unit vector in the direction $a_{ij}n_j^*$ and for a generalized longitudinal wave we put $a_i=a(t)\lambda_i$. Then we let ξ_i be the unit vector in the direction $\kappa_{ij}n_j^*$. Then (10.18) shows that across the generalized longitudinal acceleration wave $\mathscr{S}$, the heat flux amplitude is in the direction of ξ_i.

We may derive the wavespeed relation from equations (10.16) - (10.18) by multiplying (10.16) by λ_i and (10.18) by n_i, where now we denote n_i^* by n_i. Then we find

$$Q_in_i=\frac{\kappa_{ij}n_in_j}{\tau U_N}\mathscr{T},$$

and also using (10.17) we derive

$$\begin{aligned}&(\delta\tau U_N^2-\kappa_{ij}n_in_i)\mathscr{T}-a_{ij}n_j\lambda_i\tau U_Na=0,\\&(\rho U_N^2\delta_{ij}-Q_{ij})\lambda_i\lambda_ja-a_{ij}n_j\lambda_iU_N\mathscr{T}=0.\end{aligned} \tag{10.19}$$

For non-zero wave amplitudes we thus derive the wavespeed equation in the form

$$(\rho U_N^2-Q_{ij}\lambda_i\lambda_j)(\delta\tau U_N^2-\kappa_{ij}n_in_j)=(a_{ij}n_j\lambda_i)^2\tau U_N^2.$$

If we identify U_M^2 and U_T^2 as

$$U_M^2 = \frac{Q_{ij}\lambda_i\lambda_j}{\rho} \qquad \text{and} \qquad U_T^2 = \frac{\kappa_{ij}n_in_j}{\delta\tau},$$

then the wavespeed equation may be rewritten as

$$(U_N^2 - U_M^2)(U_N^2 - U_T^2) = K^2 U_N^2, \tag{10.20}$$

where

$$K^2 = \frac{(a_{ij}n_j\lambda_i)^2}{\rho\delta}.$$

The quantities U_M and U_T are the speed of a mechanical wave in the absence of thermal effects and the wavespeed of a thermal wave in the absence of mechanical effects.

From (10.20) we may assert there are two (right and left moving) waves with speeds $0 < U_2 < U_1$ such that

$$U_2^2 < \{U_T^2, U_M^2\} < U_1^2.$$

The faster wave is usually the mechanical wave whereas the slower one is normally believed to be a thermal wave.

One may proceed to calculate the amplitudes $u_i, \mathscr{T}, Q_i, P$ and Q, see exercise 10.2.

10.2 Double Voids and Type II Thermoelasticity

A type II thermoelasticity for a single distribution of voids was developed by De Cicco & Diaco [63] using the thermodynamics of Green & Naghdi [98], as described by Straughan [203], pp. 65-67. In this section we develop a type II theory of thermoelasticity with a double void structure by combining the ideas of Green & Naghdi [98] and those of Iesan & Quintanilla [115].

We begin by introducing the thermal displacement variable α of Green & Naghdi [98]

$$\alpha = \int_{t_0}^{t} \theta(\mathbf{X}, s)\,ds + \alpha_0, \tag{10.21}$$

where $\mathbf{X}$ is the spatial coordinate in the reference configuration of the body with θ being the absolute temperature. A general procedure for deriving the equations for a continuous body from a single balance of energy equation is developed by Green & Naghdi [99]. These writers derive the conservation equations for balance of mass, momentum, and entropy.

It is worth observing that Green & Naghdi [98] write, ... "*This type of theory, ... thermoelasticity type II, since it involves no dissipation of energy is perhaps a more natural candidate for its identification as thermoelasticity than the usual theory.*" Moreover, Green & Naghdi [98] observe that, ... "*This suggests that a full thermoe-*

lasticity theory - along with the usual mechanical aspects - should more logically include the present type of heat flow (type II) instead of the heat flow by conduction (classical theory, type I)." (The words in brackets have been added for clarity.) We argue that it is beneficial to develop a fully nonlinear acceleration wave analysis for a type II thermoelastic theory with double voids.

The starting point is to consider the momentum and balance of voids equations for an elastic material containing a double distribution of voids, see (3.55), (3.56) and (3.57),

$$\rho\ddot{x}_i = \pi_{Ai,A} + \rho f_i, \tag{10.22}$$

$$\kappa_1\ddot{\nu} = h_{A,A} + g + \rho\ell_1, \tag{10.23}$$

$$\kappa_2\ddot{\omega} = j_{A,A} + h + \rho\ell_2. \tag{10.24}$$

One needs a balance of energy equation and we take this to be

$$\rho\dot{\varepsilon} = \pi_{Ai}\dot{x}_{i,A} + h_A\dot{\nu}_{,A} - g\dot{\nu} + j_A\dot{\omega}_{,A} - h\dot{\omega} + \rho s\theta + (\theta\Phi_A)_{,A}. \tag{10.25}$$

In these equations X_A denote reference coordinates, x_i denote spatial coordinates, a superposed dot denotes material time differentiation and $_{,A}$ stands for $\partial/\partial X_A$. The variables $\rho, \nu, \omega, \varepsilon, \kappa_1, \kappa_2$ are the reference density, the macro void fraction, the micro void fraction, the specific internal energy, and the void inertia coefficients. The terms f_i, ℓ_1, ℓ_2 and s denote externally supplied body force, extrinsic equilibrated body forces, and externally supplied heat. The tensor π_{Ai} is the stress per unit area of the X_A-plane in the reference configuration acting over corresponding surfaces at time t (the Piola-Kirchoff stress tensor), Φ_A is the entropy flux vector, and h_A, j_A and g, h are vector and scalar functions arising in the conservation laws for voids evolution. The terms h_A and g are referred to by Nunziato & Cowin [165] in the single voids case as the equilibrated stress and the intrinsic equilibrated body force, respectively.

The next step is to use the entropy balance equation, see Green & Naghdi [98],

$$\rho\theta\dot{\eta} = \rho\theta s + \rho\theta\xi + (\theta\Phi_A)_{,A} - \Phi_A\theta_{,A} \tag{10.26}$$

where ξ is the internal rate of production of entropy per unit mass, and η, θ are the specific entropy and the absolute temperature. Introduce the Helmholtz free energy function $\psi = \varepsilon - \eta\theta$ and then equation (10.25) is rewritten with the aid of (10.26) as

$$\rho\dot{\psi} + \rho\eta\dot{\theta} = \pi_{Ai}\dot{x}_{i,A} + h_A\dot{\nu}_{,A} - g\dot{\nu} + j_A\dot{\omega}_{,A} - h\dot{\omega} + \Phi_A\theta_{,A} - \rho\theta\xi. \tag{10.27}$$

The constitutive theory we use writes the functions

$$\psi, \eta, \pi_{Ai}, \Phi_A, h_A, j_A, g, h, \xi, \tag{10.28}$$

to depend on

$$F_{iA}, \nu, \nu_{,A}, \omega, \omega_{,A}, \dot{\alpha}, \alpha_{,A}. \tag{10.29}$$

The function ψ is expanded using the chain rule, and rearranging terms, recollecting $\dot{\alpha} = \theta$, equation (10.27) may be written as

$$\begin{aligned}
&\dot{x}_{i,A}\Big(\rho\frac{\partial\psi}{\partial x_{i,A}} - \pi_{Ai}\Big) + \dot{\nu}_{,A}\Big(\rho\frac{\partial\psi}{\partial \nu_{,A}} - h_A\Big) + \dot{\omega}_{,A}\Big(\rho\frac{\partial\psi}{\partial \omega_{,A}} - j_A\Big)\\
&+\dot{\alpha}_{,A}\Big(\rho\frac{\partial\psi}{\partial \alpha_{,A}} - \Phi_A\Big) + \rho\ddot{\alpha}\Big(\frac{\partial\psi}{\partial\dot{\alpha}} + \eta\Big)\\
&+\dot{\nu}\Big(\rho\frac{\partial\psi}{\partial\nu} + g\Big) + \dot{\omega}\Big(\rho\frac{\partial\psi}{\partial\omega} + h\Big) + \rho\theta\xi = 0.
\end{aligned} \tag{10.30}$$

We now use the fact that $\dot{x}_{i,A}, \dot{\nu}_{,A}, \dot{\omega}_{,A}, \dot{\alpha}_{,A}, \ddot{\alpha}, \dot{\nu}$ and $\dot{\omega}$ appear linearly in (10.30) and so one derives the forms,

$$\begin{aligned}
&\pi_{Ai} = \rho\frac{\partial\psi}{\partial x_{i,A}}, \qquad \Phi_A = \rho\frac{\partial\psi}{\partial\alpha_{,A}}, \qquad \eta = -\frac{\partial\psi}{\partial\theta} = -\frac{\partial\psi}{\partial\dot{\alpha}},\\
&h_A = \rho\frac{\partial\psi}{\partial\nu_{,A}}, \qquad j_A = \rho\frac{\partial\psi}{\partial\omega_{,A}}, \qquad g = -\rho\frac{\partial\psi}{\partial\nu},\\
&h = -\rho\frac{\partial\psi}{\partial\omega}, \qquad \xi = 0.
\end{aligned} \tag{10.31}$$

10.2.1 Acceleration waves

We define an acceleration wave for equations (10.22) – (10.25) to be a singular surface $\mathscr{S}$ across which x_i, ν, ω and α together with their first derivatives are continuous, but the second and higher derivatives suffer a finite discontinuity. The wave amplitudes a_i, B, C and D are

$$a_i = [\ddot{x}_i], \qquad B = [\ddot{\nu}], \qquad C = [\ddot{\alpha}], \qquad D = [\ddot{\omega}].$$

By expanding equations (10.22) – (10.25) in terms of the constitutive variables and taking jumps of the resulting equations one finds, with f_i, ℓ_1, ℓ_2 and s zero, and use of the Hadamard relation (8.17)

$$\begin{aligned}
\rho a_i =& \frac{\partial\pi_{Ai}}{\partial F_{jB}}\frac{1}{U_N^2}N_AN_Ba_j + \frac{1}{U_N^2}\frac{\partial\pi_{Ai}}{\partial\nu_{,B}}N_AN_BB + \frac{1}{U_N^2}\frac{\partial\pi_{Ai}}{\partial\omega_{,B}}N_AN_BD\\
&-\frac{1}{U_N}\frac{\partial\pi_{Ai}}{\partial\dot{\alpha}}N_AC + \frac{1}{U_N^2}N_AN_B\frac{\partial\pi_{Ai}}{\partial\alpha_{,B}}C,
\end{aligned} \tag{10.32}$$

and

$$\kappa_1 B = \frac{\partial h_A}{\partial F_{iB}} \frac{1}{U_N^2} N_A N_B a_i + \frac{1}{U_N^2} \frac{\partial h_A}{\partial \nu_{,B}} N_A N_B B + \frac{1}{U_N^2} \frac{\partial h_A}{\partial \omega_{,B}} N_A N_B D \\ - \frac{1}{U_N} \frac{\partial h_A}{\partial \dot{\alpha}} N_A C + \frac{1}{U_N^2} N_A N_B \frac{\partial h_A}{\partial \alpha_{,B}} C, \tag{10.33}$$

and

$$\kappa_2 D = \frac{\partial j_A}{\partial F_{iB}} \frac{1}{U_N^2} N_A N_B a_i + \frac{1}{U_N^2} \frac{\partial j_A}{\partial \nu_{,B}} N_A N_B B + \frac{1}{U_N^2} \frac{\partial j_A}{\partial \omega_{,B}} N_A N_B D \\ - \frac{1}{U_N} \frac{\partial j_A}{\partial \dot{\alpha}} N_A C + \frac{1}{U_N^2} N_A N_B \frac{\partial j_A}{\partial \alpha_{,B}} C, \tag{10.34}$$

together with

$$-\rho \frac{\partial \varepsilon}{\partial F_{iA}} \frac{1}{U_N} a_i N_A - \rho \frac{\partial \varepsilon}{\partial \nu_{,A}} \frac{1}{U_N} N_A B - \rho \frac{\partial \varepsilon}{\partial \omega_{,A}} \frac{1}{U_N} N_A D + \rho \frac{\partial \varepsilon}{\partial \dot{\alpha}} C \\ - \rho \frac{\partial \varepsilon}{\partial \alpha_{,A}} \frac{1}{U_N} N_A C = \theta \frac{\partial \Phi_A}{\partial F_{iB}} \frac{1}{U_N^2} N_A N_B a_i + \theta \frac{\partial \Phi_A}{\partial \nu_{,B}} \frac{1}{U_N^2} N_A N_B B \\ + \theta \frac{\partial \Phi_A}{\partial \omega_{,B}} \frac{1}{U_N^2} N_A N_B D - \theta \frac{\partial \Phi_A}{\partial \dot{\alpha}} \frac{1}{U_N} N_A C + \theta \frac{\partial \Phi_A}{\partial \alpha_{,B}} \frac{1}{U_N^2} N_A N_B C \\ - \pi_{Ai} \frac{1}{U_N} N_A a_i - h_A \frac{1}{U_N} N_A B - j_A \frac{1}{U_N} N_A D, \tag{10.35}$$

where U_N is the corresponding speed of $\mathscr{S}$ at point (X_A, t) in the reference configuration.

We now examine the novel effects associated with the current theory by supposing the acceleration wave is advancing into an equilibrium region for which $\nu^+, \omega^+, \alpha^+$ and F_{iA}^+ are constants, and additionally the body has a centre of symmetry.

In this case equations (10.32) – (10.35) become,

$$(Q_{ij} - \rho U_N^2 \delta_{ij}) a_j = U_N N_A \frac{\partial \pi_{Ai}}{\partial \dot{\alpha}} C, \tag{10.36}$$

and

$$\left(\kappa_1 U_N^2 - N_A N_B \frac{\partial h_A}{\partial \nu_{,B}} \right) B = N_A N_B \frac{\partial h_A}{\partial \alpha_{,B}} C + N_A N_B \frac{\partial h_A}{\partial \omega_{,B}} D, \tag{10.37}$$

and

$$\left(\kappa_2 U_N^2 - N_A N_B \frac{\partial j_A}{\partial \omega_{,B}} \right) D = N_A N_B \frac{\partial j_A}{\partial \alpha_{,B}} C + N_A N_B \frac{\partial j_A}{\partial \nu_{,B}} B, \tag{10.38}$$

together with

$$\left(\rho\,\frac{\partial\varepsilon}{\partial\dot{\alpha}}\,U_N^2-\theta\,N_AN_B\,\frac{\partial\Phi_A}{\partial\alpha_{,B}}\right)C$$
$$=\theta\frac{\partial\Phi_A}{\partial\nu_{,B}}N_AN_BB+\theta\frac{\partial\Phi_A}{\partial\omega_{,B}}N_AN_BD+\rho U_NN_A\theta\,\frac{\partial\eta}{\partial F_{iA}}\,a_i, \tag{10.39}$$

where Q_{ij} is the acoustic tensor defined by

$$Q_{ij}=N_AN_B\,\frac{\partial\pi_{Ai}}{\partial F_{jB}}. \tag{10.40}$$

Let the position of the acceleration wave be denoted by s in the current configuration and $\mathscr{S}$ in the reference configuration. Then we may relate the normals n_i and N_A to s and $\mathscr{S}$ by the relation, see Chen [34], equation (4.10), and (8.32),

$$N_A=F_{iA}n_i\,\frac{|\nabla_{\mathbf{x}}s|}{|\nabla_{\mathbf{X}}\mathscr{S}|},$$

and then we rewrite equation (10.36) as

$$\big(Q_{ij}(n)-\rho U_N^2\delta_{ij}\big)a_j=U_N\beta_{ij}n_jC,$$

where $Q_{ij}(n)$ denotes Q_{ij} as a function of n_i rather than N_A and where the tensor β_{ij} is given by

$$\beta_{ij}=\frac{|\nabla_{\mathbf{x}}s|}{|\nabla_{\mathbf{X}}\mathscr{S}|}\,\frac{\partial\pi_{Ai}}{\partial\dot{\alpha}}\,F_{jA}\,.$$

We now use theorem 2 of Chadwick & Currie [33] to show that there is at least one direction n_j^* where $\beta_{ij}n_j^*$ is an eigenvector of Q_{ij}. The wave is propagating into an isothermal homogeneous region at rest and so the matrix Q_{ij} is constant and, therefore, $\beta_{ij}n_j^*$ is a fixed direction. We may thus consider a plane acceleration wave propagating in the direction of n_j^* but where the amplitude is in the direction of $\beta_{ij}n_j^*$. Then we let ξ_i be the unit vector in the direction $\beta_{ij}n_j^*$ and put $a_i=A\xi_i$.

One now forms the inner product of (10.36) with ξ_i. This procedure yields a system of equations in A,B,C and D. This system of equations having a zero determinant then leads to the following eighth order equation for the wavespeed U_N,

$$\begin{aligned}&(U_M^2-U_N^2)(U_N^2-U_P^2)(U_N^2-U_Q^2)(U_N^2-U_T^2)\\&-K_1^2U_N^2\{(U_N^2-U_P^2)(U_N^2-U_Q^2)-K_4^2\}\\&+K_3^2(U_M^2-U_N^2)(U_N^2-U_P^2)+K_2^2(U_M^2-U_N^2)(U_N^2-U_Q^2)\\&-K_4^2(U_N^2-U_T^2)(U_M^2-U_N^2)+2K_2K_3K_4(U_M^2-U_N^2)=0.\end{aligned} \tag{10.41}$$

In (10.41) the coefficients U_M^2, U_P^2, U_Q^2, U_T^2 are given in terms of the Helmholtz free energy by

$$U_M^2=\frac{Q_{ij}\xi_i\xi_j}{\rho}=N_AN_B\xi_i\xi_j\,\frac{\partial^2\psi}{\partial F_{iA}\partial F_{jB}}, \tag{10.42}$$

and

$$U_P^2 = \frac{N_A N_B \rho}{\kappa_1} \frac{\partial^2 \psi}{\partial \nu_{,A} \nu_{,B}}, \tag{10.43}$$

and

$$U_Q^2 = \frac{N_A N_B \rho}{\kappa_2} \frac{\partial^2 \psi}{\partial \omega_{,A} \omega_{,B}}, \tag{10.44}$$

together with

$$U_T^2 = \frac{-N_A N_B}{\psi_{\dot\alpha\dot\alpha}} \frac{\partial^2 \psi}{\partial \alpha_{,A} \partial \alpha_{,B}}. \tag{10.45}$$

Define k_1, k_2 as $k_1 = \kappa_1/\rho$ and $k_2 = \kappa_2/\rho$ and then the coefficients $K_1^2, \ldots, K_4^2$ follow from the expressions

$$\psi_{\theta\theta} K_1^2 = \left(N_A \frac{\partial^2 \psi}{\partial \theta \partial F_{iA}} \xi_i \right)^2, \tag{10.46}$$

and

$$k_1 \psi_{\theta\theta} K_2^2 = \left(N_A N_B \frac{\partial^2 \psi}{\partial \nu_{,A} \partial \alpha_{,B}} \right)^2, \tag{10.47}$$

and

$$k_2 \psi_{\theta\theta} K_3^2 = \left(N_A N_B \frac{\partial^2 \psi}{\partial \omega_{,A} \partial \alpha_{,B}} \right)^2, \tag{10.48}$$

together with

$$k_1 k_2 K_4^2 = \left(N_A N_B \frac{\partial^2 \psi}{\partial \omega_{,B} \partial \nu_{,A}} \right)^2. \tag{10.49}$$

The quantities U_M, U_P, U_Q and U_T have a physical interpretation, as now explained. The term U_M is the wavespeed of an elastic wave in the absence of other effects, U_P is the wavespeed of a wave connected to the macro void fraction, U_Q is the wavespeed of a wave connected to the micro void fraction, and U_T is the wavespeed of a thermal displacement wave. In principle equation (10.41) allows for the possibility of four distinct waves each moving to the right and the left. These four waves arise essentially as an elastic wave, a temperature wave, a wave associated to the macro voids, and a wave associated to the micro voids.

One may calculate the amplitudes A, B, C, D as functions of time, although the calculation is somewhat long.

10.2.2 Linearized Theory

We shall now develop the equations of section 10.2 for a linearized theory. Before doing this we reduce the energy balance equation (10.25). This equation is

$$\rho \dot\varepsilon = \Pi_{Ai} \dot x_{i,A} + h_A \dot\nu_{,A} - g \dot\nu + j_A \dot\omega_{,A} - h \dot\omega + (\theta \Phi_A)_{,A}. \tag{10.50}$$

We write $\varepsilon = \psi + \eta\theta$ and then we use the fact that ψ is a function of $u_{i,A}$, $\nu, \nu_{,A}$, $\omega, \omega_{,A}$, $\dot{\alpha} = \theta$ and $\alpha_{,A}$ to write the left hand side of (10.50) as

$$\begin{aligned} \rho\dot{\varepsilon} =& \rho\eta\dot{\theta} + \rho\theta\dot{\eta} + \rho(\psi_{F_{iA}}\dot{x}_{i,A} + \psi_\nu\dot{\nu} + \psi_{\nu_{,A}}\dot{\nu}_{,A} \\ &+ \psi_\omega\dot{\omega} + \psi_{\omega_{,A}}\dot{\omega}_{,A} + \psi_\theta\ddot{\alpha} + \psi_{\alpha_{,A}}\dot{\alpha}_{,A}). \end{aligned} \tag{10.51}$$

We now equate (10.51) with the right hand side of equation (10.50) and use the relations (10.31) to reduce the energy equation to

$$\theta\rho\dot{\eta} = \theta\Phi_{A,A}$$

and if we regard θ as absolute temperature then $\theta > 0$ and so the energy equation becomes

$$\rho\dot{\eta} = \Phi_{A,A}\,. \tag{10.52}$$

To obtain a linear theory we write equations (10.22), (10.23), (10.24) and (10.52) in terms of the current configuration so that with the source terms set equal to zero we find

$$\begin{aligned} \rho\ddot{u}_i &= t_{ji,j}\,, \\ \kappa_1\ddot{\nu} &= h_{i,i} + g\,, \\ \kappa_2\ddot{\omega} &= j_{i,i} + h\,, \\ \rho\dot{\eta} &= \Phi_{i,i}\,, \end{aligned} \tag{10.53}$$

where t_{ij} is the Cauchy stress tensor and h_i, j_i and Φ_i are the counterparts of h_A, J_A and Φ_A in the current configuration.

For a linearized theory we write the Helmholtz free energy as a function of the variables $e_{ij} = (u_{i,j} + u_{j,i})/2$, $\nu, \nu_{,i}$, $\omega, \omega_{,i}$, $\dot{\alpha} = \theta$ and $\alpha_{,i}$ as

$$\begin{aligned} 2\rho\psi =& a_{ijkh}e_{ij}e_{kh} + 2b_{ij}e_{ij}\nu + 2c_{ij}e_{ij}\omega + 2d_{ij}e_{ij}\theta \\ &+ f_{ij}\nu_{,i}\nu_{,j} + 2g_{ij}\nu_{,i}\omega_{,j} + h_{ij}\omega_{,i}\omega_{,j} + 2k_{ij}\nu_{,i}\alpha_{,j} \\ &+ 2\ell_{ij}\omega_{,i}\alpha_{,j} + m_{ij}\alpha_{,i}\alpha_{,j} + a_1\nu^2 + a_2\omega^2 + a_3\theta^2 \\ &+ 2a_4\nu\omega + 2a_5\nu\theta + 2a_6\omega\theta. \end{aligned} \tag{10.54}$$

If we now employ the equivalents of (10.31) in the current configuration then from (10.54) we may obtain

$$\begin{aligned} t_{ij} &= a_{ijkh}e_{kh} + b_{ij}\nu + c_{ij}\omega + d_{ij}\theta\,, \\ \Phi_i &= k_{ij}\nu_{,j} + \ell_{ij}\omega_{,j} + m_{ij}\alpha_{,j}\,, \\ \rho\eta &= -a_3\theta - a_5\nu - a_6\omega - d_{ij}e_{ij}\,, \\ h_i &= f_{ij}\nu_{,j} + g_{ij}\omega_{,j} + k_{ij}\alpha_{,j}\,, \\ j_i &= g_{ij}\nu_{,j} + h_{ij}\omega_{,j} + \ell_{ij}\alpha_{,j}\,, \\ g &= -b_{ij}e_{ij} - a_1\nu - a_4\omega - a_5\theta\,, \\ h &= -c_{ij}e_{ij} - a_2\omega - a_4\nu - a_6\theta\,. \end{aligned} \tag{10.55}$$

Upon employing (10.55) in equations (10.53) we obtain the evolution equations governing the behaviour of a thermoelastic body with a double voids distribution which is of type II thermodynamics, namely

$$
\begin{aligned}
-a_3\ddot{\alpha} &= d_{ij}\dot{u}_{i,j} + a_5\dot{\nu} + a_6\dot{\omega} + (k_{ij}\nu_{,j})_{,i} + (\ell_{ij}\omega_{,j})_{,i} + (m_{ij}\alpha_{,j})_{,i}\,,\\
\rho\ddot{u}_i &= (a_{ijkh}u_{k,h})_{,j} + (b_{ij}\nu)_{,j} + (c_{ij}\omega)_{,j} + (d_{ij}\theta)_{,j}\,,\\
\kappa_1\ddot{\nu} &= (f_{ij}\nu_{,j})_{,i} + (g_{ij}\omega_{,j})_{,i} + (k_{ij}\alpha_{,j})_{,i} - b_{ij}u_{i,j} - a_1\nu - a_4\omega - a_5\theta\,,\\
\kappa_2\ddot{\omega} &= (g_{ij}\nu_{,j})_{,i} + (h_{ij}\omega_{,j})_{,i} + (\ell_{ij}\alpha_{,j})_{,i} - c_{ij}u_{i,j} - a_2\omega - a_4\nu - a_6\theta\,.
\end{aligned}
\tag{10.56}
$$

10.2.3 Continuous Dependence Upon the Initial Data

To establish continuous dependence upon the initial data (and uniqueness) for a solution to (10.56) we suppose these equations are defined on a bounded domain Ω for $t > 0$ and the solution is subject to the boundary conditions

$$
\begin{aligned}
&u_i(\mathbf{x},t) = u_i^B(\mathbf{x},t), \qquad \nu(\mathbf{x},t) = \nu^B(\mathbf{x},t),\\
&\omega(\mathbf{x},t) = \omega^B(\mathbf{x},t), \qquad \alpha(\mathbf{x},t) = \alpha^B(\mathbf{x},t), \qquad \text{on } \Gamma\times\{t>0\},
\end{aligned}
\tag{10.57}
$$

and the initial conditions

$$
\begin{aligned}
u_i(\mathbf{x},0) &= v_i(\mathbf{x}), & \dot{u}_i(\mathbf{x},0) &= w_i(\mathbf{x}),\\
\nu(\mathbf{x},0) &= a_1(\mathbf{x}), & \dot{\nu}(\mathbf{x},0) &= a_2(\mathbf{x}),\\
\omega(\mathbf{x},0) &= b_1(\mathbf{x}), & \dot{\omega}(\mathbf{x},0) &= b_2(\mathbf{x}),\\
\alpha(\mathbf{x},0) &= c_1(\mathbf{x}), & \theta(\mathbf{x},0) &= c_2(\mathbf{x}), \qquad \mathbf{x}\in\Omega.
\end{aligned}
\tag{10.58}
$$

Denote the boundary-initial value problem comprised of (10.56), (10.57) and (10.58) by $\mathscr{P}$.

Let $(u_i^1,\nu^1,\omega^1,\alpha^1)$ and $(u_i^2,\nu^2,\omega^2,\alpha^2)$ be two solutions to $\mathscr{P}$ for the same functions u_i^B,ν^B,ω^B and α^B and let the difference solution (u_i,ν,ω,α) be defined by

$$
u_i = u_i^1 - u_i^2, \qquad \nu = \nu^1 - \nu^2, \qquad \omega = \omega^1 - \omega^2, \qquad \alpha = \alpha^1 - \alpha^2.
$$

Then (u_i,ν,ω,α) satisfies $\mathscr{P}$ with the boundary conditins being zero but with non-zero initial conditions. Note that (u_i,ν,ω,α) satisfy the equations (10.56). Mulitply $(10.56)_{1-4}$ by $\dot{\alpha} = \theta$, $\dot{u}_i$, $\dot{\nu}$ and $\dot{\omega}$ and integrate each resulting equation over Ω. Add the results and after some integration by parts and an integration in time one may arrive at the energy equation

$$
E(t) = E(0)\,,
\tag{10.59}
$$

where $E(t)$ is given by

$$\begin{aligned}E(t) =& \frac{1}{2} < \rho \dot{u}_i \dot{u}_i > + \frac{1}{2} < a_{ijkh} u_{i,j} u_{k,h} > + < b_{ij} \nu u_{i,j} > + < c_{ij} \omega u_{i,j} > \\ &+ \frac{1}{2} < \kappa_1 \dot{\nu}^2 > + \frac{1}{2} < f_{ij} \nu_{,i} \nu_{,j} > + \frac{1}{2} < a_1 \nu^2 > + \frac{1}{2} < \kappa_2 \omega^2 > \\ &+ \frac{1}{2} < h_{ij} \omega_{,i} \omega_{,j} > + \frac{1}{2} < a_2 \omega^2 > + < g_{ij} \nu_{,j} \omega_{,i} > + < k_{ij} \alpha_{,i} \nu_{,j} > \\ &+ < \ell_{ij} \alpha_{,i} \omega_{,j} > + < a_4 \omega \nu > - \frac{1}{2} < a_3 \theta^2 > + \frac{1}{2} < m_{ij} \alpha_{,i} \alpha_{,j} > .\end{aligned}$$

We suppose ρ, κ_1, κ_2 and $-a_3$ are strictly positive everywhere and $E(t)$ defines a positive-definite form and then continuous dependence upon the initial data follows from (10.59).

10.3 Exercises

Exercise 10.1. Consider the displacement boundary-initial value problem for equations (10.9). Under the conditions of section 10.1.1 establish uniqueness and continuous dependence upon the initial data for a solution to this displacement boundary-initial value problem.

Exercise 10.2. For a one-dimensional wave moving into an equilibrium region, calculate the amplitude equation for an acceleration wave as defined in section 10.1.2. Solve this equation to find the amplitudes $a, P, Q, \mathscr{T}$ and $\hat{Q}$, where $\hat{Q}$ is the jump in the time derivative of the one-dimensional heat flux.

Exercise 10.3. Define an acceleration wave for equations (10.4). Using an analysis like that of section 10.1.2 find the wavespeeds of a generalized longitudinal wave.

Exercise 10.4. For a one-dimensional acceleration wave for exercise 10.3 moving into an equilibrium region, calculate and solve the amplitude equation.

Exercise 10.5. Define an acceleration wave for equations (10.56). Find the wavespeed equations for such a wave and interpret. Find and solve the amplitude equation for a one-dimensional acceleration wave moving into an equilibrium region.

References

[1] Aguilera, R.F., Aguilera, R.: A triple - porosity model for petrophysical analysis of naturally fractured reservoirs. Petrophysics **45**, 157–166 (2004)

[2] Alharbi, A.M., Scott, N.H.: Wave stability in generalized temperature - rate - dependent thermoelasticity. IMA J. Appl. Math. **81**, 750–778 (2016)

[3] Alharbi, A.M., Scott, N.H.: Stability in constrained temperature - rate - dependent thermoelasticity. Mathematics and Mechanics of Solids **22**, 1738–1763 (2017)

[4] Amendola, G., Fabrizio, M.: Thermomechanics of damage and fatigue by a phase field model (2014). ArXiv: 1410.7042

[5] Aouadi, M., Lazzari, B., Nibbi, R.: A theroy of thermoelasticity with diffusion under Green - Naghdi models. Z. Angew. Math. Mech. **94**, 837–852 (2014)

[6] Arbogast, T., Douglas, J., Hornung, U.: Derivation of the double porosity model of single phase flow via homogenization theory. SIAM J. Math. Anal. **21**, 823–836 (1990)

[7] Avalishvili, G., Avalishvili, M., Müller, W.H.: On the well-posedness of the Green-Lindsay model. Mathematics and Mechanics of Complex Systems **5**, 115–125 (2017)

[8] Bai, M., Elsworth, D., Roegiers, J.C.: Modelling of naturally fractured reservoirs using deformation dependent flow mechanism. Int. J. Rock Mechanics and Mining Sciences **30**, 1185–1191 (1993)

[9] Bai, M., Elsworth, D., Roegiers, J.C.: Multiporosity/multipermeability approach to the simulation of naturally fractured reservoirs. Water Resources Research **29**, 1621–1633 (1993)

[10] Bai, M., Roegiers, J.C.: Triple - porosity analysis of solute transport. J. Contaminant Hydrology **28**, 247–266 (1997)

[11] Barbosa, A., Rafael, A., Ma, T.F.: Long - time dynamics of an extensible plate equation with thermal memory. J. Math. Anal. Appl. **416**, 143–165 (2014)

[12] Barenblatt, G.I., Zheltov, Y.P.: Fundamental equations of filtration of homogeneous liquids in fissured rock. Soviet Physics Doklady **5**, 522–525 (1960)

B. Straughan, *Mathematical Aspects of Multi–Porosity Continua*, Advances in Mechanics and Mathematics 38, https://doi.org/10.1007/978-3-319-70172-1

[13] Barenblatt, G.I., Zheltov, Y.P., Kochina, I.N.: Basic concepts in the theory of seepage of homogeneous liquids in fissured rocks (strata). J. Appl. Math. Mech. **24**, 1286–1303 (1960)

[14] Berryman, J.G., Wang, H.F.: The elastic coefficients of double - porosity models for fluid transport in jointed rock. J. Geophys. Res. B **100**, 24,611–24,627 (1995)

[15] Berryman, J.G., Wang, H.F.: Elastic wave propagation and attenuation in a double - porosity dual - permeability medium. Int. J. Rock Mechanics and Mining Sciences **37**, 63–78 (2000)

[16] Berti, A., Bochicchio, I., Fabrizio, M.: Phase separation in quasi-incompressible fluids: Cahn - Hilliard model in the Cattaneo - Maxwell framework. ZAMP **66**, 135–147 (2015)

[17] Berti, A., Fabrizio, M., Giorgi, C.: A three - dimensional phase transition model in ferromagnetism: existence and uniqueness. J. Math. Anal. Appl. **355**, 661–674 (2009)

[18] Biot, M.A.: Theory of propagation of elastic waves in a fluid-saturated porous solid. I. Low frequency range. J. Acous. Soc. America **28**, 168–178 (1956)

[19] Biot, M.A.: Theory of finite deformations of porous solids. Indiana Univ. Math. J. **21**, 597–620 (1972)

[20] Bissell, J.J.: Oscillatory convection with the Cattaneo - Christov hyperbolic heat flow model. Proc. Roy. Soc. London A **471**, 20140,845 (2015)

[21] Bissell, J.J.: Thermal convection in a magnetized conducting fluid with the Cattaneo - Christov heat flow model. Proc. Roy. Soc. London A **472**, 20160,649 (2016)

[22] Bonnetti, E., Colli, P.L., Fabrizio, M., Gilardi, G.: Existence of solutions for a mathematical model related to solid - solid phase transitions is shape memory alloys. Arch. Rational Mech. Anal. **291**, 203–254 (2016)

[23] Borchardt, C.W.: Über die transformation der Elasticitätsgleichungen in allgemeine orthogonale Koordinaten. J. Reine Angew. Math. **76**, 45–58 (1873)

[24] Borja, R.L., Liu, X., White, J.A.: Multiphysics hillslope processes triggering landslides. Acta Geotechnica **7**, 261–269 (2012)

[25] Borja, R.L., White, J.A.: Continuum deformation and stability analyses of a steep hillside slope under rainfall infiltration. Acta Geotechnica **5**, 1–14 (2010)

[26] Both, S., Czel, B., Fulop, T., Grof, G., Gyenis, A., Kovacs, R., Van, P., Verhas, J.: Deviation from the Fourier law in room - temperature heat pulse experiments. J. Non-Equilibrium Thermodynamics **41**, 41–48 (2016)

[27] Capelli, A., Kapil, J.C., Reiweger, I., Or, D., Schweizer, J.: Speed and attenuation of acoustic waves in snow: Laboratory experiments and modelling with Biot's theory. Cold Regions Sci. Tech. **125**, 1–11 (2016)

[28] Caputo, M., Fabrizio, M.: Damage and fatigue described by a fractional derivative model. J. Computational Physics **293**, 400–408 (2015)

[29] Cariou, S., Dormieux, L., Skoczylas, F.: An original constitutive law for Calloro-Oxfordian argillite, a two scale double-porosity material. Applied Clay Science **81**, 18–30 (2013)

[30] Carneiro, J.F.: Numerical simulations on the influence of matrix diffusion to carbon sequestration in double porosity fissured aquifers. Int. J. Greenhouse Gas Control **3**, 431–443 (2009)

[31] Cavaterra, C., Grasselli, M., Wu, H.: Non-isothermal viscous Cahn - Hilliard equation with inertial terms and dynamic boundary conditions. Comm. Pure Appl. Anal. **13**, 1855–1890 (2014)

[32] Caviglia, G., Morro, A., Straughan, B.: Thermoelasticity at cryogenic temperatures. Intl. J. Nonlinear Mech. **27**, 251–263 (1992)

[33] Chadwick, P., Currie, P.K.: Intrinsically characterized acceleration waves in heat - conducting elastic materials. Proc. Camb. Phil. Soc. **76**, 481–491 (1974)

[34] Chen, P.J.: Growth and decay of waves in solids. In: S. Flügge, C. Truesdell (eds.) Handbuch der Physik, vol. VIa/3, pp. 303–402. Springer, Berlin (1973)

[35] Chen, Z.: Homogenization and simulation for compositional flow in naturally fractured reservoirs. J. Math. Anal. Appl. **326**, 12–32 (2007)

[36] Chen, Z.X.: Transient flow of slightly compressible fluids through double porosity, double permeability systems - a state of the art review. Trans. Porous Media **4**, 147–184 (1989)

[37] Chirita, S.: On the strong ellipticity condition for transversely isotropic linearly elastic solids. An. St. Univ. Iasi Matematica **52**, 245–250 (2006)

[38] Chirita, S.: Rayleigh waves on an exponentially graded poroelastic half space. J. Elasticity **110**, 185–199 (2013)

[39] Chirita, S., Ciarletta, M., Straughan, B.: Structural stability in porous elasticity. Proc. Roy. Soc. London A **462**, 2593–2605 (2006)

[40] Chirita, S., Ciarletta, M., Tibullo, V.: Rayleigh surface waves on a Kelvin-Voigt viscoelastic half space. J. Elasticity **115**, 61–76 (2014)

[41] Chirita, S., Ciarletta, M., Tibullo, V.: On the wave propagation in the time differential dual - phase - lag thermal model. Proc. Roy. Soc. London A **471**, 20150,400 (2015)

[42] Chirita, S., Danescu, A.: Strong ellipticity for tetragonal system in linearly elastic solids. Int. J. Solids Structures **45**, 4850–4859 (2008)

[43] Chirita, S., Danescu, A., Ciarletta, M.: On the strong ellipticity of the anisotropic linearly elastic materials. J. Elasticity **87**, 1–27 (2007)

[44] Chirita, S., Ghiba, I.D.: Strong ellipticity and progressive waves in elastic materials with voids. Proc. Roy. Soc. London A **466**, 439–458 (2010)

[45] Choo, J., White, J.A., Borja, R.I.: Hydromechanical modeling of unsaturated flow in double porosity media. Int. J. Geomechanics **16**, D4016,002 (2016)

[46] Christov, I.C.: Nonlinear acoustics and shock formation in lossless barotropic Green-Naghdi fluids. Evolutionary Equations and Control Theory **5**, 349–365 (2016)

[47] Christov, I.C., Christov, C.I., Jordan, P.M.: Modelling weakly nonlinear acoustic wave propagation. Ql. J. Appl. Math. Mech. **60**, 473–495 (2007)

[48] Christov, I.C., Jordan, P.M.: On the propagation of second sound in nonlinear media: shock, acceleration and travelling wave results. J. Thermal Stresses **33**, 1109–1135 (2010)

[49] Christov, I.C., Jordan, P.M., Chin-Bing, S.A., Warn-Varnas, A.: Acoustic travelling waves in thermoviscous perfect gases: Kinks, acceleration waves, and shocks under the Taylor-Lighthill balance. Mathematics and Computers in Simulation **127**, 2–18 (2016)

[50] Ciarletta, M., Iesan, D.: Non-classical elastic solids. Longman, New York (1993)

[51] Ciarletta, M., Passarella, F., Svanadze, M.: Plane waves and uniqueness theorems in the coupled linear theory of elasticity for solids with double porosity. J. Elasticity **114**, 55–68 (2014)

[52] Ciarletta, M., Straughan, B.: Thermo-poroacoustic acceleration waves in elastic materials with voids. J. Math. Anal. Appl. **333**, 142–150 (2007)

[53] Ciarletta, M., Straughan, B., Zampoli, V.: Thermo-poroacoustic acceleration waves in elastic materials with voids without energy dissipation. Int. J. Engng. Sci. **45**, 736–743 (2007)

[54] Coleman, B.D., Noll, W.: The thermodynamics of elastic materials with heat conduction and viscosity. Arch. Rational Mech. Anal. **13**, 167–178 (1963)

[55] Cook, J.D., Showalter, R.E.: Microstructure diffusion models with secondary flux. J. Math. Anal. Appl. **189**, 731–756 (1995)

[56] Coti Zelati, M., Dell'Oro, F., Pata, V.: Energy decay of type III linear thermoelastic plates with memory. J. Math. Anal. Appl. **401**, 357–366 (2013)

[57] Cowin, S.C.: The viscoelastic behaviour of linear elastic materials with voids. J. Elasticity **15**, 185–191 (1985)

[58] Dai, W.Z., Kuang, Z.B.: Love waves in double porosity media. J. Sound and Vibration **296**, 1000–1012 (2006)

[59] Dai, W.Z., Kuang, Z.B., Zhao, S.X.: Rayleigh waves in a double porosity half-space. J. Sound and Vibration **298**, 319–332 (2006)

[60] D'Apice, C., Chirita, S.: Plane harmonic waves in the theory of thermoviscoelastic materials with voids. J. Thermal Stresses **39**, 142–155 (2016)

[61] David, C., Menéndez, B., Darot, M.: Influence of stress - induced and thermal cracking on physical properties and microstructure of La Peyrette granite. Int. J. Rock Mechanics and Mining Sciences **36**, 433–448 (1999)

[62] Dazel, O., Bécot, F.X., Jaouen, L.: Biot effects for sound absorbing double porosity materials. Acta Acustica united with Acustica **98**, 567–576 (2012)

[63] De Cicco, S., Diaco, M.: A theory of thermoelastic materials with voids without energy dissipation. J. Thermal Stresses **25**, 493–503 (2002)

[64] De Sciarra, F.M., Salerno, M.: On thermodynamic functions in thermoelasticity without energy dissipation. European J. Mech. A - Solids **46**, 84–95 (2014)

[65] Debois, G., Urai, J.L., Hemes, S., Schröppel, B., Schwarz, J.O., Mac, M., Weiel, D.: Multi - scale analysis of porosity in diagenetically altered reservoir sandstone from Permian Rotliegand (Germany). J. Pet. Sci. Engng. **140**, 128–148 (2016)

[66] Dejaco, A., Komlev, V.S., Jaroszewicz, J., Swieszkowski, W., Hellmich, C.: Micro CT-based multiscale elasticity of double - porous(pre-cracked) hy-

droxyapatite granules for regenerative medicine. J. Biomechanics **45**, 1068–1075 (2012)
[67] Dejaco, A., Komlev, V.S., Jaroszewicz, J., Swieszkowski, W., Hellmich, C.: Fracture safety of double - porous hydroxyapatite biomaterials. Bioinspired Biomimetric and Nanobiomaterials **5**, 24–36 (2016)
[68] Dell'Oro, F., Pata, V.: Memory relaxation of type III thermoelastic extensible beams and Berger plates. Evolution Equations and Control Theory **1**, 251–270 (2012)
[69] Deng, J.H., Leguizamon, J.A., Aguilera, R.: Petrophysics of triple - porosity tight gas reservoirs with a link to gas productivity. SPE Reservoir Evaluation and Engineering **14**, SPE–144,590–PA (2011)
[70] Dufresne, M., Bacchin, P., Cerino, G., Remigy, J.C., Adrianus, G.N., Aimar, P., Legallais, C.: Human hepatic cell behaviour on polysulfone membrane with double porosity level. J. Membrane Science **428**, 454–461 (2013)
[71] Duwairi, H.M.: Sound wave propagation in porous layers. J. Porous Media **17**, 723–730 (2014)
[72] Eltayeb, I.: Stability of a porous Bénard - Brinkman layer in local thermal non-equilibrium with Cattaneo effects in solid. Int. J. Thermal Science **98**, 208–218 (2015)
[73] Enterria, M., Suarez-Garcia, F., Martinez-Alonso, A., Tascon, J.M.D.: Preparation of hierarchical micro-mesoporous aluminosilicate composites by simple Y zeolite / MCM-48 silica assembly. J. Alloys and Compounds **583**, 60–69 (2014)
[74] Fabrizio, M.: Some remarks on the fractional Cattaneo - Maxwell equation for heat propagation. Fractional Calculus and Applied Analysis **214**, 1074–1079 (2006)
[75] Fabrizio, M., Franchi, F.: Delayed thermal models: stability and thermodynamics. J. Thermal Stresses **37**, 160–173 (2014)
[76] Fabrizio, M., Franchi, F., Straughan, B.: On a model for thermo-poroacoustic waves. Int. J. Engng. Sci. **46**, 790–798 (2008)
[77] Fabrizio, M., Giorgi, C., Morro, A.: Non-isothermal phase-field approach to the second sound transition in solids. Il Nuovo Cimento B: General Physics, Relativity, Astronomy and Mathematical Physics and Methods **121**, 383–399 (2006)
[78] Fabrizio, M., Giorgi, C., Morro, A.: A thermodynamic approach to non-isothermal phase-field evolution in continuum physics. Physica D - Nonlinear Phenomena **214**, 144–156 (2006)
[79] Fabrizio, M., Lazzari, B.: Stability and second law of thermodynamics in dual - phase - lag heat conduction. Int. J. Heat Mass Transfer **74**, 484–489 (2014)
[80] Fatori, L.H., Silva, M.A.J., Ma, T.F., Yang, Z.J.: Long - time behaviour of a class of thermoelastic plates with nonlinear strain. J. Differential Equations **259**, 4831–4862 (2015)
[81] Fellah, Z.E.A., Fellah, M., Depollier, C.: Transient acoustic wave propagation in porous media. Tech. rep., Intech (2016). Http://dx.doi.org/10.5772/55048

[82] Fernandez, J.R., Masid, M.: Analysis of a problem arising in porous thermoelasticity of type II. J. Thermal Stresses **39**, 513–531 (2016)
[83] Fitzpatrick, C.M.: Nonlinear corrections to Darcy's law at low Reynolds numbers. Ph.D. thesis, University College London (2011)
[84] Flavin, J.N.: Thermoelastic Rayleigh waves in a prestressed medium. Proc. Camb. Phil. Soc. **58**, 532–538 (1962)
[85] Flavin, J.N., Green, A.E.: Plane thermoelastic waves in an initially stressed medium. J. Mech. Phys. Solids **9**, 179–190 (1961)
[86] Franchi, F., Lazzari, B., Nibbi, R.: Mathematical models for the non-isothermal Johnson - Segelman viscoelasticity in porous media: stability and wave propagation. Math. Meth. Appl. Sci. **38**, 4075–4087 (2015)
[87] Gabitto, J., Tsouris, C.: Modelling the capacitive deionization process in dual - porosity electrodes. Trans. Porous Media. **113**, 173–205 (2016)
[88] Galdi, G., Rionero, S.: Continuous dependence theorems in linear elasticity on exterior domains. Int. J. Engng. Sci. **17**, 521–526 (1979)
[89] Gelet, R., Loret, B., Khalili, N.: Borehole stability analysis in a thermo-poroelastic dual-porosity medium. Int. J. Rock Mech. Mining Sci. **50**, 65–76 (2012)
[90] Gentile, M., Straughan, B.: Acceleration waves in nonlinear double porosity elasticity. Int. J. Engng. Sci. **73**, 10–16 (2013)
[91] Ghasemizadeh, R., Hellweger, F., Butscher, C., Padilla, I., Vesper, D., Field, M., Alshawabkeh, A.: Review: Groundwater flow and transport modelling of karst aquifers, with particular reference to the North Coast Limestone aquifer system of Puerto Rico. Hydrogeol. J. **20**, 1441–1461 (2012)
[92] Gilbarg, D., Trudinger, N.S.: Elliptic Partial Differential Equations of Second Order. Springer, Heidelberg (1977)
[93] Giorgi, C., Grandi, D., Pata, V.: On the Green - Naghdi type III heat conduction model. Discrete and Continuous Dynamical Systems B **19**, 2133–2143 (2014)
[94] Goodman, M.A., Cowin, S.C.: A continuum theory for granular materials. Arch. Rational Mech. Anal. **44**, 249–266 (1972)
[95] Gorgas, T.J., Wilkens, R.H., Fu, S.S., Frazer, L.N., Richardson, M.D., Briggs, K.B., Lee, H.: In situ acoustic and laboratory ultrasonic sound speed and attenuation measured in heterogeneous soft seabed sediments: Eel River shelf, California. Marine Geology **182**, 103–119 (2002)
[96] Greaves, G.N., Greer, A.L., Lakes, R.S., Rouxel, T.: Poisson's ratio and modern materials. Nature Materials **10**, 823–837 (2011)
[97] Green, A.E., Naghdi, P.M.: A re-examination of the basic postulates of thermomechanics. Proc. Roy. Soc. London A **432**, 171–194 (1991)
[98] Green, A.E., Naghdi, P.M.: Thermoelasticity without energy-dissipation. J. Elasticity **31**, 189–208 (1993)
[99] Green, A.E., Naghdi, P.M.: A unified procedure for construction of theories of deformable media. I. Classical continuum physics. Proc. Roy. Soc. London A **448**, 335–356 (1995)

[100] Ha, C.S., Hestekin, E., Li, J., Plesha, M.E., Lakes, R.S.: Controllable thermal expansion of large magnitude in chiral negative Poisson's ratio lattices. Physica Status Solidi b **252**, 1431–1434 (2015)
[101] Hayat, T., Muhammad, T., Al-Mezal, S., Liao, S.J.: Darcy-Forchheimer flow with variable thermal conductivity and Cattaneo-Christov heat flux. Int. J. Num. Meth. Heat Fluid Flow **26**, 2355–2369 (2016)
[102] Hayat, T., Zubair, M., Ayub, M., Waqas, M., Alsaedi, A.: Stagnation point flow towards nonlinear stretching surface with Cattaneo - Christov heat flux. European Physical Journal Plus **131**, 1–10 (2016)
[103] He, J., Teng, W., Xu, J., Jiang, R., Sun, J.: A quadruple - porosity model for shale gas reservoirs with multiple migration mechanisms. J. Natural Gas Science and Engineering **33**, 918–933 (2016)
[104] Hirsch, M.W., Smale, S.: Differential equations, dynamical systems, and linear algebra. Academic Press, New York (1974)
[105] Homand-Etienne, F., Houpert, R.: Thermally induced microcracking in granites: characterization and analysis. Int. J. Rock Mechanics and Mining Sciences **26**, 125–134 (1989)
[106] Hornung, U., Showalter, R.E.: Diffusion models for fractured media. J. Math. Anal. Appl. **147**, 69–80 (1990)
[107] Huang, Y.G., Shirota, Y., Wu, M.Y., Su, S.Q., Yao, Z.S., Kang, S., Kanegawa, S., Li, G.L., Wu, S.Q., Kamachi, T., Yoshizawa, K., Ariga, K., Hong, M.C., Sata, O.: Superior thermoelasticity and shape-memory nanopores in a porous supramolecular organic framework. Nature Communications **7**, 11,564 (2016)
[108] Iesan, D.: Incremental equations in thermoelasticity. J. Thermal Stresses **3**, 41–56 (1980)
[109] Iesan, D.: A theory of thermoelastic materials with voids. Acta Mechanica **60**, 67–89 (1986)
[110] Iesan, D.: Thermoelasticity of initially heated bodies. J. Thermal Stresses **11**, 17–38 (1988)
[111] Iesan, D.: Thermoelastic Models of Continua. Kluwer, Dordrecht (2004)
[112] Iesan, D.: Second-order effects in the torsion of elastic materials with voids. ZAMM **85**, 351–365 (2005)
[113] Iesan, D.: Nonlinear plane strain of elastic materials with voids. Math. Mech. Solids **11**, 361–384 (2006)
[114] Iesan, D.: On a theory of thermoviscoelastic materials with voids. J. Elasticity pp. 369–384 (2011)
[115] Iesan, D., Quintanilla, R.: On a theory of thermoelastic materials with a double porosity structure. J. Thermal Stresses **37**, 1017–1036 (2014)
[116] Iesan, D., Quintanilla, R.: Strain gradient theory of chiral Cosserat thermoelasticity without energy dissipation. J. Math. Anal. Appl. **437**, 1219–1235 (2016)
[117] Iesan, D., Scalia, A.: Thermoelastic Deformations. Kluwer, Dordrecht (1996)

[118] Jiang, H.L., Dou, Y.H., Xi, Z.C., Chen, M., Jin, Y.: Microscopic choked flow in a highly compressible gas in porous media. J. Natural Gas Science and Engineering **35**, 42–53 (2016)

[119] John, F.: Continuous dependence on data for solutions of partial differential equations with a prescribed bound. Comm. Pure Appl. Math. **13**, 551–585 (1960)

[120] Jordan, P.M.: Growth, decay and bifurcation of shock amplitudes under the type-II flux law. Proc. Roy. Soc. London A **463**, 2783–2798 (2007)

[121] Jordan, P.M.: Some remarks on nonlinear poroacoustic phenomena. Mathematics and Computers in Simulation **80**, 202–211 (2009)

[122] Jordan, P.M.: A note on poroacoustic waves. Physics Letters A **377**, 1350–1357 (2013)

[123] Jordan, P.M.: Second - sound phenomena in inviscid, thermally relaxing gases. Discrete and Continuous Dynamical Systems B **19**, 2189–2205 (2014)

[124] Jordan, P.M.: The effects of coupling on finite amplitude acoustic travelling waves in thermoviscous gases: Blackstock's models. Evolutionary Equations and Control Theory **5**, 383–397 (2016)

[125] Jordan, P.M.: A survey of weakly nonlinear acoustic models, 1910–2009. Mech. Res. Comm. **73**, 127–139 (2016)

[126] Jordan, P.M., Fulford, J.K.: A note on poroacoustic travelling waves under Darcy's law: exact solutions. Applications of Mathematics **50**, 99–115 (2011)

[127] Jordan, P.M., Passarella, F., Tibullo, V.: Poroacoustic waves under a mixture - theoretic based reformulation of the Jordan - Darcy - Cattaneo model. Wave Motion **71**, 82–92 (2017)

[128] Jou, D., Carlomagno, I., Cimmelli, V.A.: A thermodynamic model for heat transport and thermal wave propagation in graded systems. Physica E: Low - Dimensional Systems and Nanostructures **73**, 242–249 (2015)

[129] Jou, D., Carlomagno, I., Cimmelli, V.A.: Rectification of low-frequency thermal waves in graded SicGe1-c. Physics Letters A **380**, 1824–1829 (2016)

[130] Jou, D., Casas-Vázquez, J., Lebon, G.: Extended Irreversible Thermodynamics, fourth edn. Springer, Berlin (2010)

[131] Jou, D., Cimmelli, V.A.: Constitutive equations for heat conduction in nanosystems and nonequilibrium processes: an overview. Comm. Appl. Industrial Math. **7**, 196–222 (2016)

[132] Kaltenbacher, B.: Mathematics of nonlinear acoustics. Evolution Equations and Control Theory **4**, 447–491 (2015)

[133] Khalili, N.: Coupling effects in double porosity media with deformable matrix. Geophys. Res. Letters **30**, 2153 (2003)

[134] Khalili, N., Selvadurai, A.P.S.: A fully coupled constitutive model for thermo-hydro-mechanical analysis in elastic media with double porosity. Geophys. Res. Letters **24**, 2268 (2003)

[135] Khayat, R.E., deBruyn, J., Niknami, M., Stranges, D.F., Khorasany, R.M.H.: Non-Fourier effects in macro- and micro-scale non-isothermal flow of liquids and gases: review. Int. J. Thermal Sci. **97**, 163–177 (2015)

[136] Kim, J., Moridis, G.J.: Numerical analysis of fracture propagation during hydraulic fracturing operations in shale gas systems. Int. J. Rock Mechanics and Mining Sciences **76**, 127–137 (2015)

[137] Kim, S., Hosseini, S.A.: Hydro - thermo - mechanical analysis during injection of cold fluid into a geologic formation. Int. J. Rock Mechanics and Mining Sciences **77**, 220–236 (2015)

[138] Knops, R.J., Payne, L.E.: Stability in linear elasticity. Int. J. Solids Structures **4**, 1233–1242 (1968)

[139] Knops, R.J., Payne, L.E.: Stability of the traction boundary value problem in linear elastodynamics. Int. J. Engng. Sci. **4**, 351–357 (1968)

[140] Knops, R.J., Payne, L.E.: Uniqueness in classical elastodynamics. Arch. Rational Mech. Anal. **27**, 349–355 (1968)

[141] Knops, R.J., Payne, L.E.: Continuous data dependence for the equations of classical elastodynamics. Proc. Camb. Phil. Soc. **66**, 481–491 (1969)

[142] Knops, R.J., Payne, L.E.: Growth estimates for solutions of evolutionary equations in Hilbert space with applications in elastodynamics. Arch. Rational Mech. Anal. **41**, 363–398 (1971)

[143] Knops, R.J., Payne, L.E.: Uniqueness theorems in linear elasticity, *Tracts in Natural Philosophy*, vol. 19. Springer, Berlin, Heidelberg (1971)

[144] Knops, R.J., Payne, L.E.: Improved estimates for continuous data dependence in linear elastodynamics. Math. Proc. Camb. Phil. Soc. **103**, 535–559 (1988)

[145] Knops, R.J., Wilkes, E.W.: Theory of elastic stability. In: C. Truesdell (ed.) Handbuch der Physik, pp. 125–302. Springer, Berlin (1973). Volume VIa/3

[146] Konyukhov, A., Pankratov, L.: New non-equilibrium matrix imbibition equation for double porosity model. Comptes Rendus Mécanique **344**, 510–520 (2016)

[147] Kuang, Z.B.: Energy principles for temperature varied with time. Int. J. Thermal Sci. **120**, 80–85 (2017)

[148] Kuznetsov, A.V., Nield, D.A.: The onset of convection in a tridisperse porous medium. Int. J. Heat Mass Transfer **54**, 3120–3217 (2011)

[149] Lakes, R.S.: Elastic and viscoelastic behaviour of chiral materials. Int. J. Mech. Sciences **43**, 1579–1589 (2001)

[150] Lebeau, G., Zuazua, E.: Decay rates for the three-dimensional linear systems of thermoelasticity. Arch. Rational Mech. Anal. **148**, 179–231 (1999)

[151] Leseduarte, M.C., Quintanilla, R.: On uniqueness and continuous dependence in type III thermoelasticity. J. Math. Anal. Appl. **395**, 429–436 (2012)

[152] Lindsay, K.A., Straughan, B.: Acceleration waves and second sound in a perfect fluid. Arch. Rational Mech. Anal. **68**, 53–87 (1978)

[153] Liu, L., Zheng, L.C., Liu, F.W., Zhang, X.X.: Anomalous convection diffusion and wave coupling transport of cells on comb frame with fractional Cattaneo - Christov flux. Comm. Nonlinear Sci. Numerical Simulation **38**, 45–58 (2016)

[154] Liu, L., Zheng, L.C., Liu, F.W., Zhang, X.X.: An imposed heat conduction model with Riesz fractional Cattaneo - Christov flux. Int. J. Heat Mass Transfer **103**, 1191–1197 (2016)

[155] Liu, Y.: Structural stability results for thermoelasticity of type III. Bull. Korean Math. Soc. **51**, 1269–1279 (2014)

[156] Ly, H.B., Droumaguet, B.L., Monchiet, V., Grande, D.: Facile fabrication of doubly porous polymeric materials with controlled nano- and micro-porosity. Polymer **78**, 13–21 (2015)

[157] Marin, M., Nicaise, S.: Existence and stability results for thermoelastic dipolar bodies with double porosity. Cont. Mech. Thermodyn. **28**, 1645–1657 (2016)

[158] Masin, D., Herbstova, V., Bohac, J.: Properties of double porosity clayfills and suitable constitutive models (2012). http://web.natur.cuni.cz/uhigug/masin/download/mhb_16ICSMGE_Osaka.pdf

[159] Masters, I., Pao, W.K.S., Lewis, R.W.: Coupling temperature to a double - porosity model of deformable porous media. Int. J. Numerical Methods in Engineering **49**, 421–438 (2000)

[160] Miller, W., Ren, Z., Smith, C.W., Evans, K.E.: A negative Poisson's ratio carbon fibre composite using a negative Poisson's ratio yarn reinforcement. Composites Science and Technology **72**, 761–766 (2012)

[161] Molinari, V., Mostacci, D., Ganapol, B.D.: Wave propagation in an ideal gas: first and second sound. J. Comp. Thoer. Transport **45**, 268–274 (2016)

[162] Montrasio, L., Valentino, R., Losi, G.L.: Rainfall infiltration in a shallow soil: a numerical simulation of the double - porosity effect. Electronic J. Geotechnical Engineering **16**, 1387–1403 (2011)

[163] Mouhat, F., Coudert, F.X.: Necessary and sufficient elastic stability conditions in various crystal systems. Phys. Rev. B **90**, 224,104 (2014)

[164] Mustafa, M., Hayat, T., Alsaedi, A.: Rotating flow of Maxwell fluid with variable thermal conductivity: an application to non-Fourier heat flux theory. Int. J. Heat Mass Transfer **106**, 142–148 (2017)

[165] Nunziato, J.W., Cowin, S.C.: A nonlinear theory of elastic materials with voids. Arch. Rational Mech. Anal. **72**, 175–201 (1979)

[166] Nunziato, J.W., Walsh, E.K.: On the influence of void compaction and material non-uniformity on the propagation of one-dimensional acceleration waves in granular materials. Arch. Rational Mech. Anal. **64**, 299–316 (1977)

[167] Nunziato, J.W., Walsh, E.K.: Addendum, On the influence of void compaction and material non-uniformity on the propagation of one-dimensional acceleration waves in granular materials. Arch. Rational Mech. Anal. **67**, 395–398 (1978)

[168] Ogden, R.W.: Non-linear Elastic Deformations. Dover, Mineola, New York (1997)

[169] Olusola, B.K., Yu, G., Aguilera, R.: The use of electromagnetic mixing rules for petrophysical evaluation of dual- and triple-porosity reservoirs. SPE Reservoir Evaluation and Engineering **16**, 378–389 (2013)

[170] Payne, L.E.: Isoperimetric inequalities and their applications. SIAM Review **9**, 453–488 (1967)

[171] Payne, L.E., Polya, G., Weinberger, H.F.: On the ratio of consecutive eigenvalues. Stud. Appl. Math. **35**, 289–298 (1956)

[172] Payne, L.E., Weinberger, H.F.: New bounds for solutions of second order elliptic partial differential equations. Pacific J. Math. **8**, 551–573 (1958)

[173] Payne, L.E., Weinberger, H.F.: An optimal Poincaré inequality for convex domains. Arch. Rational Mech. Anal. **5**, 182–188 (1960)

[174] Pooley, E.J.: Centrifuge modelling of ground improvement for double porosity clay. Dissertation No. 21647, ETH Zurich, Institut für Geotechnik (2015)

[175] Quintanilla, R.: On the linear problem of swelling porous elasic solids. J. Math. Anal. Appl. **269**, 50–72 (2002)

[176] Quintanilla, R., Sivaloganathan, J.: Aspects of the nonlinear theory of type II thermostatics. European J. Mech. A - Solids **32**, 109–117 (2012)

[177] Randrianalisoa, J.H., Dombrovsky, L.A., Lipinski, W., Timchenko, V.: Effects of short - pulsed laser radiation on transient heating of superficial human tissues. Int. J. Heat Mass Transfer **78**, 488–497 (2014)

[178] Rionero, S., Galdi, G.P.: On the uniqueness of viscous fluid motions. Arch. Rational Mech. Anal. **62**, 295–301 (1976)

[179] Rohan, E., Naili, S., Nguyen, U.H.: Wave propagation in a strongly heterogeneous elastic porous medium: homogenization of Biot medium with double porosities. Comptes Rendus Mécanique **344**, 569–581 (2016)

[180] Rossmanith, D., Puri, A.: Nonlinear evolution of a sinusoidal pulse under a Brinkman - based poroacoustic model. Int. J. Nonlinear Mech. **78**, 53–58 (2016)

[181] Rossmanith, D., Puri, A.: Recasting a Brinkman - based acoustic model as the damped Burgers equation. Evolution Equations and Control Theory **5**, 463–474 (2016)

[182] Rukolaine, S.: Unphysical effects of the dual - phase - lag model of heat conduction. Int. J. Heat Mass Transfer **78**, 58–63 (2014)

[183] Sabelnikov, V.A., Petrova, N.N., Lipatnikov, A.N.: Analytical and numerical study of travelling waves using the Maxwell - Cattaneo relaxation model extended to reaction - advection - diffusion systems. Phys. Rev. E **94**, 042,218 (2016)

[184] Said, B., Grandjean, A., Barre, Y., Tancret, F., Fajula, F., Galameau, A.: LTA zeolite monoliths with hierarchical trimodal porosity as highly efficient microreactors for strontium capture in continuous flow. Microporous amd Mesoporous Materials **232**, 39–52 (2016)

[185] Sakamoto, M., Matsumoto, T.: Development and evaluation of superporous ceramics bone tissue scaffold materials with triple pore structure a) Hydroxyapatite b) Beta - tricalcium phosphate. In: H. Tal (ed.) Bone regeneration. InTech, Croatia (2012). Cdn.intechopen.com/pdfs-wm/34836.pdf

[186] Sanami, M., Ravirala, N., Alderson, K., Alderson, A.: Auxetic materials for sports applications. Procedia Engineering **72**, 453–458 (2014)

[187] Scotto di Santolo, A., Evangelista, A.: Calibration of a rheological model for debris flow hazard mitigation in the Campania region. In: Z. Chen, J.M. Zhang, K. Ho, F.Q. Wu, Z.K. Li (eds.) Landslides and engineered slopes. From the past to the future, pp. 913–919. Taylor and Francis, London (2008)

[188] Sarma, P., Aziz, K.: New transfer functions for simulation of naturally fractured reservoirs with dual - porosity models. SPE J. **50**, 328–340 (2006)

[189] Scarpetta, E., Svanadze, M.: Uniqueness theorems in the quasi - static theory of thermoelasticity for solids with double porosity. J. Elasticity **120**, 67–86 (2015)

[190] Scarpetta, E., Svanadze, M., Zampoli, V.: Fundamental solutions in the theory of thermoelasticity for solids with double porosity. J. Thermal Stresses **37**, 727–748 (2014)

[191] Sellitto, A., Rogolino, P., Carlomagno, I.: Heat-pulse propagation along nonequilibrium nanowires in thermomass theory. Comm. Appl. Industrial Math. **7**, 39–55 (2016)

[192] Sgard, F., Olny, X., Atalla, N., Castel, F.: On the use of perforations to improve the sound absorption of porous materials. Applied Acoustics **66**, 625–651 (2005)

[193] Sharma, M.D.: Rayleigh waves in a partially saturated poroelastic solid. Geophys. J. Int. **189**, 1203–1214 (2012)

[194] Showalter, R.E.: Diffusion in poro-elastic media. J. Math. Anal. Appl. **251**, 310–340 (2000)

[195] Showalter, R.E., Visarraga, D.B.: Double-diffusion models from a highly heterogeneous medium. J. Math. Anal. Appl. **295**, 191–210 (2004)

[196] Showalter, R.E., Walkington, N.J.: Micro-structure models of diffusion in fissured media. J. Math. Anal. Appl. **155**, 1–20 (1991)

[197] Singh, A., Tadmor, E.B.: Thermal parameter identification for non - Fourier heat transfer from molecular dynamics. J. Computational Physics **299**, 667–686 (2015)

[198] Solano, N.A., Clarkson, C.R., Krause, F.F., Aquino, S.D., Wiseman, A.: On the characterization of unconventional oil reservoirs. Canadian Society of Exploration Geophysicists Recorder **38**, 42–47 (2013)

[199] Spencer, A.J.M.: Continuum Mechanics. Longman, London (1980)

[200] Spicer, J.B., Olasov, L.R., Zeng, F.W., Han, K., Gallego, N.C., Contescu, C.I.: Laser ultrasonic assessment of the effects of porosity and microcracking on the elastic moduli of nuclear graphites. J. Nuclear Materials **471**, 80–91 (2016)

[201] Straughan, B.: Qualitative analysis of some equations in contemporary continuum mechanics. Ph.D. thesis, Heriot-Watt University (1974)

[202] Straughan, B.: Stability and Wave Motion in Porous Media, *Appl. Math. Sci.*, vol. 165. Springer, New York (2008)

[203] Straughan, B.: Heat Waves, *Appl. Math. Sci.*, vol. 177. Springer, New York (2011)

[204] Straughan, B.: Stability and uniqueness in double porosity elasticity. Int. J. Engng. Science **65**, 1–8 (2013)

[205] Straughan, B.: Shocks and acceleration waves in modern continuum mechanics and in social systems. Evolution Equations and Control Theory **3**, 541–555 (2014)

[206] Straughan, B.: Convection with local thermal non-equilibrium and microfluidic effects, *Advances in Mechanics and Mathematics*, vol. 32. Springer, New York (2015)
[207] Straughan, B.: Modelling questions in multi - porosity elasticity. Meccanica **51**, 2957–2966 (2016)
[208] Straughan, B.: Waves and uniqueness in multi-porosity elasticity. J. Thermal Stresses **39**, 704–721 (2016)
[209] Straughan, B.: Uniqueness and stability in triple porosity thermoelasticity. Rendiconti Lincei, Matematica e Applicazioni **28**, 191–208 (2017)
[210] Sui, J., Zheng, L., Zhang, X.: Boundary layer heat transfer with Cattaneo-Christov double diffusion in upper convected Maxwell nanofluid past a stretching sheet with slip velocity. Int. J. Thermal Sciences **104**, 461–468 (2016)
[211] Svanadze, M.: Dynamical problems of the theory of elasticity for solids with double porosity. Proc. Appl. Math. Mech. **10**, 309–310 (2010)
[212] Svanadze, M.: Plane waves and boundary value problems in the theory of elasticity for solids with double porosity. Acta Applicandae Mathematicae **122**, 461–471 (2012)
[213] Svanadze, M.: On the theory of viscoelasticity for materials with double porosity. Discrete and Continuous Dynamical Systems B **19**, 2335–2352 (2014)
[214] Svanadze, M.: Uniqueness theorems in the theory of thermoelasticity for solids with double porosity. Meccanica **49**, 2099–2108 (2014)
[215] Svanadze, M.: Fundamental solutions in the theory of elasticity for triple porosity materials. Meccanica **51**, 1825–1837 (2016)
[216] Svanadze, M., Scalia, A.: Mathematical problems in the theory of bone poroelasticity. Biomath **1**, 1–4 (2012)
[217] Svanadze, M., Scalia, A.: Mathematical problems in the coupled linear theory of bone poroelasticity. Computers and Mathematics with Applications **66**, 1554–1566 (2013)
[218] Thomson, W.: Reflection and refraction of light. Phil. Mag. **26**, 414–425 (1888)
[219] Truesdell, C.: Existence of longitudinal waves. J. Acous. Soc. Amer. **40**, 729–730 (1966)
[220] Truesdell, C., Noll, W.: The Non-linear Field Theories of Mechanics, second edn. Springer, Berlin (1992)
[221] Truesdell, C., Toupin, R.A.: The classical field theories. In: S. Flügge (ed.) Handbuch der Physik, vol. III/1, pp. 226–793. Springer (1960)
[222] Van, P.: Theories and heat pulse experiments of non-Fourier heat conduction. Comm. Appl. Industrial Math. **7**, 150–166 (2016)
[223] Venegas, R., Umnova, O.: Acoustic properties of double porosity granular materials. J. Acous. Soc. America **130**, 2765–2776 (2011)
[224] Wakeni, M.F., Reddy, B.D., McBride, A.T.: An unconditionally stable algorithm for generalized thermoelasticity based on operator splitting and time -

discontinuous Galerkin finite element methods. Comp. Meth. Appl. Mech. Engng. **306**, 427–451 (2016)

[225] Warren, J.R., Root, P.J.: The behaviour of naturally fractured reservoirs. Soc. Pet. Eng. J. **228**, 245–255 (1963)

[226] Wei, Z., Zhang, D.: Coupled fluid - flow and geomechanics for triple - porosity / dual - permeability modelling of coalbed methane recovery. International Journal of Rock Mechanics and Mining Sciences **47**, 1242–1253 (2008)

[227] Winkler, K.W., Murphy, W.F.: Acoustic velocity and attenuation in porous rocks, pp. 20–34. American Geophysical Union (1995). Rock physics and phase relations, handbook of physical constants

[228] Xinchun, S., Lakes, R.S.: Stability of elastic material with negative stiffness and negative Poisson's ratio. Physica Status Solidi b **244**, 1008–1026 (2007)

[229] Yuan, J., Sundén, B.: On mechanisms and models of multi-component gas diffusion in porous structures of fuel cell electrodes. Int. J. Heat Mass Transfer **69**, 358–374 (2014)

[230] Zhao, Y., Chen, M.: Fully coupled dual - porosity model for anisotropic formations. International Journal of Rock Mechanics and Mining Sciences **43**, 1128–1133 (2006)

[231] Zhao, Y., Zhang, L., Zhao, J., Luo, J., Zhang, B.: Triple porosity modelling of transient well test and rate decline analysis for multi-fractured horizontal well in shale gas reservoirs. J. Petroleum Science and Engineering **110**, 253–262 (2013)

[232] Zhou, D., Gao, Y., Lai, M., Li, H., Yuan, B., Zhu, M.: Fabrication of NiTi shape memory alloys with graded porosity to imitate human long - bone structure. J. Bionic Engineering **12**, 575–582 (2015)

[233] Zou, M., Wei, C., Yu, H., Song, L.: Modelling and application of coalbed methane recovery performance based on a triple porosity / dual permeability model. J. Natural Gas Science and Engineering **22**, 679–688 (2015)

[234] Zuber, A., Motyka, J.: Hydraulic parameters and solute velocities in triple - porosity karstic - fissured - porous carbonate aquifers: case studies in southern Poland. Environmental Geology **34**, 243–250 (1998)

Index

B. Straughan, *Mathematical Aspects of Multi–Porosity Continua*, Advances in Mechanics and Mathematics 38, https://doi.org/10.1007/978-3-319-70172-1

Printed in the United States
By Bookmasters